POLYMER MELT PROCESSING

Most of the shaping in the manufacture of polymeric objects is carried out in the melt state, as is a substantial part of the physical property development. Melt processing involves an interplay between fluid mechanics and heat transfer in rheologically complex liquids, and taken as a whole it is a nice example of the importance of coupled transport processes. This book is about the underlying foundations of polymer melt processing, which can be derived from relatively straightforward ideas in fluid mechanics and heat transfer; the level is that of an advanced undergraduate or beginning graduate course, and the material can serve as the text for a course in polymer processing or for a second course in transport processes.

Morton M. Denn is the Albert Einstein Professor of Science and Engineering and Director of the Benjamin Levich Institute for Physico-Chemical Hydrodynamics at the City College of New York, CUNY. Prior to joining CCNY in 1999, he was Professor of Chemical Engineering at the University of California, Berkeley, where he served as Department Chair, as well as Program Leader for Polymers and Head of Materials Chemistry in the Materials Sciences Division of the Lawrence Berkeley National Laboratory. He previously taught chemical engineering at the University of Delaware, where he was the Allan P. Colburn Professor. Professor Denn was editor of the *AIChE Journal* from 1985 to 1991 and editor of the *Journal of Rheology* from 1995 to 2005. He is the recipient of a Guggenheim Fellowship; a Fulbright Lectureship; the Professional Progress, William H. Walker, Warren K. Lewis, and Institute Lectureship Awards of the American Institute of Chemical Engineers; the Chemical Engineering Lectureship of the American Society for Engineering Education; and the Bingham Medal and Distinguished Service Awards of the Society of Rheology. He is a member of the National Academy of Engineering and the American Academy of Arts and Sciences, and he received an honorary D.Sc. from the University of Minnesota. His previous books are *Optimization by Variational Methods, Introduction to Chemical Engineering Analysis* (as coauthor), *Stability of Reaction and Transport Processes, Process Fluid Mechanics,* and *Process Modeling.*

CAMBRIDGE SERIES IN CHEMICAL ENGINEERING

Books in the Series:

Polymer Melt Processing

FOUNDATIONS IN FLUID MECHANICS AND HEAT TRANSFER

Morton M. Denn

City College of New York

CAMBRIDGE
UNIVERSITY PRESS

CAMBRIDGE
UNIVERSITY PRESS

32 Avenue of the Americas, New York NY 10013-2473, USA

Cambridge University Press is part of the University of Cambridge.

It furthers the University's mission by disseminating knowledge in the pursuit of education, learning and research at the highest international levels of excellence.

www.cambridge.org
Information on this title: www.cambridge.org/9781107417496

First published 2008
First paperback edition 2014

A catalogue record for this publication is available from the British Library

Library of Congress Cataloguing in Publication data

Denn, Morton M., 1939–
Polymer melt processing : foundations in fluid mechanics and heat transfer / Morton M. Denn.
   p. cm. – (Cambridge series in chemical engineering)
Includes bibliographical references and index.
ISBN 978-0-521-89969-7 (hardback)
1. Polymer melting.   2. Fluid mechanics.   3. Heat – Transmission.
I. Title. II. Series.
TP156.P6D46   2008
668.4'13–dc22   2008004882

ISBN 978-0-521-89969-7 Hardback
ISBN 978-1-107-41749-6 Paperback

Contents

Color insert follows page 86.

Preface

Most of the shaping in the manufacture of polymeric objects is carried out in the melt state, as is a substantial part of the physical property development. Melt processing involves an interplay between fluid mechanics and heat transfer in rheologically complex liquids, and taken as a whole it is a nice example of the importance of coupled transport processes. This is a book about the underlying foundations of polymer melt processing, which can be derived from relatively straightforward ideas in fluid mechanics and heat transfer; the level is that of an advanced undergraduate or beginning graduate course, and the material can serve as the text for a course in polymer processing or for a second course in transport processes. The book is based on a course that has evolved over thirty years, which I first taught at the University of Delaware and subsequently at the University of California, Berkeley; the Hebrew University of Jerusalem; and the City College of New York. The target audience is twofold: engineers and physical scientists interested in polymer processing who seek a firm command of basic principles without getting into details of the process geometry or the fluid rheology, and students who wish to apply the basic material from courses in transport processes to practical processing situations. The only background necessary is some prior study of the fundamentals of fluid flow and heat transfer and a command of mathematics at a level typically expected of an advanced undergraduate student in engineering or the physical sciences; the text is otherwise self-contained.

The book begins with introductory material and a brief review of fundamentals, after which the first part focuses on analytical treatments of basic polymer processes: extrusion, mold filling, fiber spinning, and so forth. The thin gap (lubrication) and thin filament approximations are employed, and all analyses in this part are for inelastic liquids. An introduction to finite element calculation follows, where full numerical solutions are compared to analytical results. Polymer rheology is then introduced, with an emphasis on relatively simple viscoelastic models that have been used with some success to model processing operations. Applications in which melt viscoelasticity is important are then revisited, followed by a chapter on stability and sensitivity that focuses on melt spinning and a chapter on wall slip and extrusion instabilities. There are brief concluding chapters on structured fluids and mixing and dispersion.

The viscoelastic character of polymer melts reflects the entangled microstructure and plays an important role in property development and in flow stability. Viscoelasticity has little effect on the evolution of many processing flows, however, where the mechanics are dominated by the temperature and shear-rate dependence of the viscosity; this statement is especially true of extrusion and some mold filling, but it applies as well to some extensional flows when the polymer is a relatively inelastic polyester or nylon. I have therefore chosen to develop the subject for inelastic liquids, and only then, with the complete framework in place, to introduce the effects of elasticity where appropriate. This sequence is in contrast to the usual approach of starting with polymer melt rheology, but I believe that exploiting students' understanding of purely viscous liquids to lay the foundation is superior pedagogy.

I am grateful to Benoit Debbaut, George Vassilatos, and Kurt Wissbrun for detailed comments on every chapter. Benoit Debbaut also contributed his expertise in numerical simulation as the coauthor of Chapter 8. I am also grateful to the many authors whose work is used and cited at the appropriate places in the text, and of course to students and co-workers who have contributed so much to my understanding throughout the years. The late Arthur B. Metzner first introduced me to problems in polymer processing more than forty years ago, and I believe that his influence will be evident throughout the text. My wife, Vivienne, contributed in more ways than I can possibly express.

Morton M. Denn
New York
November 2007

1 Polymer Processing

1.1 Introduction

Polymeric materials – often called *plastics* in popular usage – are ubiquitous in modern life. Applications range from film to textile fibers to complex electronic interconnects to structural units in automobiles and airplanes to orthopedic implants. Polymers are giant molecules, consisting of hundreds or thousands of connected *monomers*, or basic chemical units; a polyethylene molecule, for example, is simply a chain of covalently bonded carbon atoms, each carbon containing two hydrogen atoms to complete the four valence sites. The polyethylene used to manufacture plastic film typically has an average molecular weight (called the *number-average molecular weight*, denoted M_n) of about 29,000, or about 2,000 $-CH_2-$ units, each with a molecular weight of 14. The symbol "$-$" on each side of the CH_2 denotes a single covalent bond with the adjacent carbon atom. (The monomer is actually ethylene, $CH_2=CH_2$, where "$=$" denotes a double bond between the carbons that opens during the polymerization process, and a single "mer" is $-CH_2-CH_2-$; hence, the molecular weight of the monomer is 28 and the degree of polymerization is about 1,000.) The *ultra-high molecular weight* polyethylene used in artificial hips and other prosthetic devices has about 36,000 $-CH_2-$ units. Polystyrene is also a chain of covalently bonded carbon atoms, but one hydrogen on every second carbon is replaced with a phenyl (benzene) ring. Two or more monomers might be polymerized together to form a *copolymer*, appearing on the chain in either a regular or random sequence. The monomers for some common engineering plastics are shown in Table 1.1.

The polymers used in commercial applications are solids at their use temperatures. The solid phase might be brittle or ductile, depending on the chemical composition and, to some extent, on the way in which the polymer has been processed. The chemical composition of the backbone of some polymers, such as polyethylene, is such that crystallization can occur; other polymers, such as polystyrene, cannot form crystalline structures and solidify only as amorphous solids, or *glasses*. The *glass transition* occurs when the temperature is sufficiently low to prevent large-scale chain motion. Crystallization and glass transition temperatures are shown for

Table 1.1. *Repeat units and transition temperatures of some common polymers*

Polymer	Monomer	$T_g(°C)$	$T_m(°C)$
Linear polyethylene (HDPE)	(–C–C–) with H H above and H H below	~ 110	134
Branched polyethylene (LDPE)	(–C–C–) with H H above and H H below	~ 110	115
Polystyrene	(–C–C–) with H H above and H φ below	90–100	none
Poly(ethylene terephthalate)	(–O–C–C–O–C–φ–C–) with H H O O substituents	70	260
Poly(methylmethacrylate)	(–C–C–) with H CH₃ above and H C–O–CH₃, ‖ O below	90–100	none

Note: φ denotes a phenyl group (a benzene ring); the substitution is in the para position in poly(ethylene terephthalate).

the polymers in Table 1.1. Polymers are very viscous in the liquid state, and molecular diffusion is slow; hence, the molecular reorganization necessary to permit crystallization can sometimes be so slow that a crystallizable polymer will reach the glass transition temperature and solidify as a glass before crystallization can occur. Indeed, there are always amorphous regions in any crystalline polymer.

Polymers are often blended or contain additives to affect the properties of the solid phase; high-impact polystyrene, for example, is a blend in which particles of a rubbery polymer, typically polybutadiene or a styrene-butadiene copolymer, are dispersed in polystyrene. Many polymer composites used in molding applications contain solid fillers, such as calcium carbonate particles, glass fibers, or even nanoscale fillers like exfoliated clays or carbon nanotubes.

The polymer manufacturer, starting from raw materials like natural gas and other low molecular weight chemicals, produces the polymer – say, polyethylene – as a powder or in the form of chips or flakes, which are often converted (densified) into pellets by extrusion. This *resin* must be processed to produce the desired product – a molded part, for example. Most processing takes place in the liquid state. The resin must first be melted (we will use the term *melt* to denote the change from any form of solid to a liquid state, although technically only a crystal has a true melting transition), then conveyed through one or more steps to form an object of the desired shape, and finally solidified again. A polymer pipe, for example, is produced by continuously extruding the molten polymer through an annular die and then cooling it quickly to retain the shape. An injection-molded part is produced by forcing the molten polymer into a mold of the desired shape, where the polymer cools until it has solidified, after which the mold is opened, the part removed, and the process repeated.

We typically draw the backbone carbons in a polymer chain in a straight line, but in fact sequential covalent bonds between carbon atoms form an angle of $109.5°$, not $180°$. Rotation about the bonds between adjacent carbon atoms permits substantial lateral motion of the chain, and any directional correlation of the backbone usually vanishes over lengths of five or ten monomer units; this distance is known as a *Kuhn length*. A single polymer molecule therefore has the appearance of a very long, flexible string of beads, or even simply a very long flexible rope, and the dynamics of isolated polymer molecules in dilute solutions are very well described by a statistical mechanics treatment of a string of beads undergoing random motion in the presence of Brownian forces. If these molecules are now packed together in a melt or a concentrated solution, it is clear that the motion of one string of beads is highly constrained by all the other strings in its neighborhood. The best visual picture is a bowl of spaghetti; if we put a fork into the bowl and attempt to move one strand, we impose a motion on all the other strands with which that strand is in contact, and we cause a macroscopic motion that tends to align the *entangled* strands. In a similar manner, the deformation we impose on a polymer melt during processing can induce orientation in the chains at a molecular level, and this orientation in turn manifests itself in the distribution of stresses, the ability of the melt to crystallize during cooling, and the mechanical and optical properties of the bulk material. The properties of the final shaped object thus depend in part on the chemical nature of the particular polymer and in part on the details of the mechanical process and the stress and thermal fields to which the melt has been subjected.

Our goal in this text is to use mathematical modeling to develop the basic principles necessary to understand polymer melt processing, to analyze and predict behavior, and ultimately to develop the tools needed to guide process operation and design. A mathematical model is an abstraction that captures the essential features of a physical process in a set of equations that can be manipulated, analytically or numerically, to explain and predict behavior. A good model is based on fundamental physical principles, with essential compromises between the detail required for fidelity to the underlying physical phenomena and the simplicity required for practical implementation. The level of the model is determined by the anticipated application. The academic tools we will employ in modeling polymer melt processes are fluid mechanics and heat transfer. The natural language of these subjects is multivariable calculus, at a level commonly taught during the second year in U.S. science and engineering curricula; because we are interested in phenomena that vary spatially in more than one direction and may vary in time, the physical phenomena will be described by partial differential equations. The actual applications, however, in

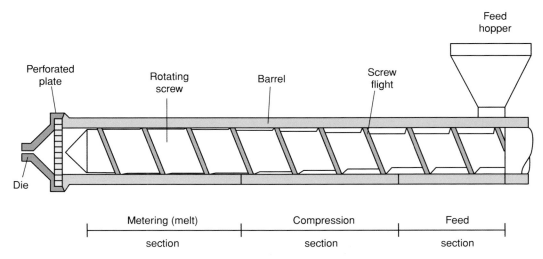

Figure 1.1. Schematic of a single-screw extruder.

contrast to the language itself and the formal structures of the models, require little or no experience in the solution of partial differential equations; indeed, simple quadratures of the type encountered in an introductory calculus course suffice for most situations that we will encounter.

1.2 Typical Processes

In the following sections we briefly describe some of the most important melt processes, and we will return to the modeling of these processes in subsequent chapters. We will be looking at generic features, and we will not focus on mechanical detail, which is of course very important in actual operation.

1.2.1 Extrusion

Extrusion is the most fundamental and most widely used unit operation in melt processing. An extruder is a device that pressurizes a melt in order to force it through a shaping die or some other unit. A ram extruder, for example, is simply a piston that forces a melt from a cylinder through a die. We are usually concerned with continuous extrusion over long periods of time, in which case a ram, which must operate in a semibatch mode (i.e., the cylinder must be refilled periodically), is not appropriate. The most common device for continuous extrusion is the single-screw extruder, shown schematically in Figure 1.1. The single-screw extruder is analogous to the meat grinder that was once a fixture in kitchens. In a meat grinder, chunks of meat are placed in a hopper and fall onto a rotating *Archimedes screw*. The meat is compressed and carried forward by the screw flights until it is forced through a perforated plate, producing the strands that make up "ground" meat. The counterintuitive feature here, which we rarely think about in the context of a meat grinder, is that the meat enters at atmospheric pressure and is forced through the perforated

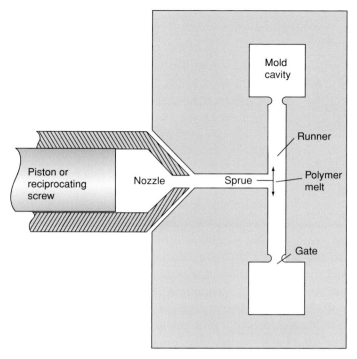

Figure 1.2. Schematic of injection molding.

plate to emerge at atmospheric pressure; hence, the pressure must *increase* from the hopper to the upstream side of the plate in order to provide the force necessary to push the meat through the plate. We are accustomed to thinking in terms of pressure *drop* in flow situations. Similarly, in a screw extruder the polymer, in the form of flakes, chips, or pellets, is fed through the hopper onto the screw, where melting takes place because of frictional and conductive heating and perhaps also deformation heating of the softening solid. The polymer is conveyed forward by the screw, becoming completely molten by the time it reaches the *metering section*. Pressure builds up in the flow direction until the end of the screw, where the polymer is forced through a shaping die. The pressure drop through the die must equal the pressure buildup along the screw.

Twin-screw extruders, in which the screws intermesh while conveying the polymer, are also in common use. Twin screws are very effective mixing devices, and they are commonly used for compounding blends and composites, as well as for reactive processes, in which a chemical reaction occurs in the extruder.

1.2.2 Injection Molding

Injection molding is a semibatch operation shown schematically in Figure 1.2. The process is conceptually very straightforward: Molten polymer is forced into a closed mold from a ram extruder or a screw extruder with a reciprocating screw and allowed to solidify, after which the mold is opened, the part is removed, and the

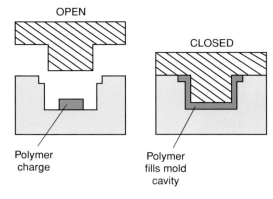

OPEN

CLOSED

Figure 1.3. Schematic of compression molding.

Polymer charge

Polymer fills mold cavity

process is repeated. There are a number of practical issues, however. The mold walls are typically cold, so the polymer is cooling during the filling cycle. If the mold is filled too slowly and too much solidification occurs, the mold cavity will be closed off before filling is complete. Indeed, incomplete filling can occur even without solidification because of a large melt viscosity increase from cooling during the filling process. Incomplete filling is a particular problem in the manufacture of complex molded parts, like those used for electronic interconnects. In addition, the flow inside the mold is very important. The flow plays a significant role in determining the morphology of the finished part, which in turn determines the physical properties. Many molds have inserts around which the polymer must flow, or multiple "gates" to facilitate filling, and the weld lines where the flow fronts meet can be mechanically weak points; hence, design to ensure optimal placement of the weld lines is an important consideration. Finally, it is common to fill more than one mold cavity from a single extruder, as indicated in the schematic. Flow balancing to ensure equal flow to all molds is therefore very important. Very high pressures can be reached in injection molding; polymer melts are usually considered to be incompressible, but this is one application where compressibility of the melt may be important because of the very high pressures. The high pressures also have implications regarding mechanical design; leakage around the mold face can be important because of inadequate pressure to keep the mold sealed, for example.

1.2.3 Compression Molding

Compression molding, shown schematically in Figure 1.3, is also conceptually simple. Polymer is placed between two mold faces and flows out to fill the cavity as the mold is closed. The charge for compression molding of large parts – an automotive hood, for example – often consists of stacked layers of *sheet molding compound,* which is a fiber-filled polymer sheet that can be handled at room temperature. The plies may be oriented in various directions to achieve desired fiber orientation in the final product, especially if the fibers are long (*continuous fibers*). When the fibers are short, the fiber orientation distribution is determined by the flow field during mold

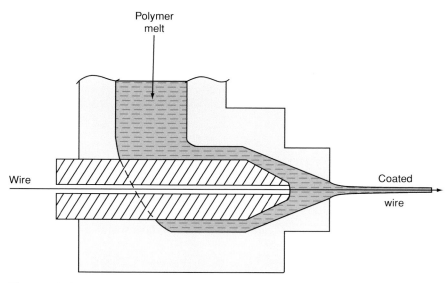

Figure 1.4. Schematic of a "cross-head" wire coating die.

closing. The programming of mold closing, especially in a large mold that might weigh several tons, is an interesting exercise; the feedback signal controlling the rate of closure can be position, force, or some combination of both. Compression molding is usually carried out with *thermosetting* polymers, which polymerize during the processing.

1.2.4 Coating

There are many types of coating operations. We will focus here on the coating of wire and film, shown schematically in Figure 1.4. A wire or film (the *substrate*) is passed through the die. The thickness of the coating for a given die geometry is determined by the substrate speed and the upstream pressure, since both parameters contribute to the polymer flow rate. Coating uniformity is an important consideration here, especially when the visual appearance of the coating is important; small variations in coating thickness in certain wavelength ranges can have a large impact on reflectivity, for example. The interior design of the die is important in order to prevent regions of melt recirculation, in which the organic polymer spends long times in the die at high temperature, since polymer degradation can occur and produce coating defects when the degraded polymer finally leaves the die. In some film extrusion coating processes, the polymer is extruded onto the moving substrate, rather than being contacted inside the die. This latter process, in which the extruded melt is stretched as it passes from the die to the sheet, is very similar to fiber spinning, which is described next. An instability known as *draw resonance*, in which the film thickness varies periodically, is a major concern in this process and one that we shall discuss subsequently.

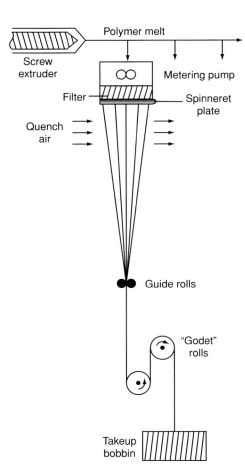

Figure 1.5. Schematic of fiber spinning.

1.2.5 Fiber Spinning

The melt spinning process for the manufacture of textile fibers is shown schematically in Figure 1.5. Let us first focus on a single filament. The polymer from the extruder, after passing through a filter to remove small gel particles, is forced through a small hole known as a *spinneret*, which is typically on the order of 200–400 μm in diameter. The jet, which might be at a temperature of 290 °C for poly(ethylene terephthalate), or PET, the polyester typically used in textile fibers, emerges into an ambient environment that is below the solidification temperature, which for poly(ethylene terephthalate) is about 80 °C. The filament is taken up on a roll moving at a much higher linear velocity than the extrusion velocity; the takeup speed is typically in excess of 3,000 m/min (50 m/s, or about 120 mph), while the average linear velocity through the spinneret is typically two orders of magnitude smaller. Mass conservation therefore requires that the filament at the point of takeup be drawn down in area by a factor roughly equal to the ratio of the takeup velocity to the extrusion velocity ("roughly" because the densities at the spinneret and takeup will be different because of the large temperature difference). The filament solidifies between the spinneret and the takeup, and the drawdown will occur mostly in the melt phase.

The mechanics of this process clearly depend on the interplay between the fluid mechanics causing melt stretching and the very high rate of air cooling; cooling affects the viscosity and hence the resistance of the fluid to stretching. The high speeds involved introduce aerodynamic considerations; air drag and filament inertia are important contributors to the filament mechanics, and the nature of the boundary layer in the air stream around the filament plays a significant role. In most cases the spinneret plate contains a large number of holes, and the individual filaments are taken up together as a *yarn*. The cooling air therefore contacts each filament in a different way, causing each filament to deform differently. Stretching flow is an efficient means of polymer chain orientation, which helps determine the final fiber morphology and properties; different stretching histories on different filaments will therefore cause some property variation within the yarn.

The primary operating concern is filament uniformity and the avoidance of breaks. The melt zone below the spinneret but prior to solidification is short, typically on the order of one meter. The residence time in the melt zone is therefore on the order of fractions of a second. Feedback control of the average filament diameter on this time scale is not feasible. Furthermore, each extruder feeds many spinning stations, and a fiber plant will contain hundreds of stations. Thus, this process essentially operates in an "open loop" mode, with operator adjustments taking place over time scales that are very long relative to the time a fluid element spends on the line. Models are very helpful in defining process operating strategies, and major process improvements have been effected with the guidance of spinline models.

Polymer films are formed in a number of ways, but one common film process looks like a two-dimensional version of the spinning process, in which a molten sheet is extruded into air and stretched, after which the film is solidified and taken up on a cold roll. Film processes tend to operate at much lower speeds than fiber processes, but the basic mechanics are the same except near the edges of the film, where three-dimensional effects are important.

1.2.6 Film Blowing

The blown film process, shown schematically in Figure 1.6, is commonly used to manufacture biaxially oriented films and plastic bags. A thin cylindrical film is extruded through an annular die. The inside pressure is slightly above ambient, causing the film to expand (like a rubber balloon). The film is flattened at "hauloff" and taken up at a linear speed higher than the linear extrusion velocity, so stretching occurs both in the "machine direction" and in the transverse direction. Solidification occurs prior to hauloff. The flattened annular film is slit on the sides if film is the desired product or processed further to form periodic seals if bags are the product. The blown film process is very sensitive to operate, and the aerodynamics around the air ring seem to have a major effect on bubble stability.

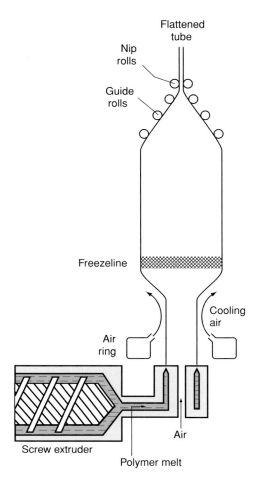

Figure 1.6. Schematic of the blown film process.

1.2.7 Blow Molding and Thermoforming

The blow molding process, which is used for the manufacture of bottles and industrial components such as automotive fuel tanks, combines elements of a number of the previous processes. A tube that is closed at one end, known as a *parison,* is first formed, either by injection molding or by extrusion. The heated parison is then pressurized and stretched to conform to the shape of the mold, where it solidifies. The blowing portion of the cycle is a biaxial stretch.

Classical thermoforming is a similar process in which a sheet is heated and deformed by vacuum from inside the mold or by pressurization from outside to stretch and conform to the shape of the mold. Some thermoforming processes utilize a mechanical device for part or all of the deformation of the sheet. Thermoforming is used to produce high-volume thin-walled products such as drinking cups and food packaging as well as large items like cargo bed liners for pickup trucks.

The modeling of the inflation and solidification portions of blow molding and thermoforming do not introduce major new concepts beyond those incorporated in

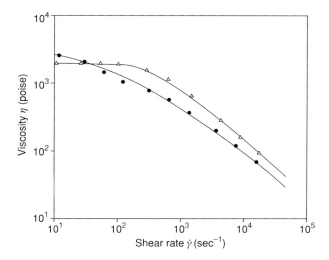

Figure 1.7. Viscosities of two polystyrenes as functions of shear rate. Open triangles: $M_w = 160{,}000$, $M_w/M_n < 1.1$; closed circles, $M_w = 261{,}000$, $M_w/M_n < 2.5$. Reprinted with permission from Graessley et al., *J. Rheol.*, **14**, 519 (1970).

other free-surface processes, and we will not address these processes separately in later chapters, despite their processing importance.

1.3 Polymer Melt Rheology

Rheology is the study of deformation and flow. Since melt processing entails a large amount of deformation and flow, we may expect melt rheology to play a significant role, and this is indeed the case. Polymer melts are *non-Newtonian* in that the relation between the stress and deformation rate is nonlinear. They are also *viscoelastic*, which means that the entangled network of polymer chains sometimes responds in a manner more reminiscent of a rubbery solid than of a liquid. We will find that we can make considerable headway in understanding polymer processing without taking the melt rheology explicitly into account, and this will be our starting point. Ultimately, of course, we will have to address the non-Newtonian, viscoelastic nature of the melts in a quantitative manner, but it is pedagogically advantageous to put that time off. For now, it suffices to introduce only a few concepts relating to melt rheology.

1.3.1 Viscosity

Viscosity is defined as the ratio of shear stress to shear rate. The viscosity of a *Newtonian fluid* is a material constant that depends on temperature and pressure but is independent of the rate of shear; that is, the shear stress is directly proportional to the shear rate at fixed temperature and pressure. Low molar-mass liquids and all gases are Newtonian. Complex liquids, such as polymers and suspensions, tend to be non-Newtonian in that the shear stress is a nonlinear function of the shear rate. Some typical melt viscosities are shown in Figure 1.7. The viscosity approaches a constant value at low shear rates, known as the *zero-shear viscosity* and denoted

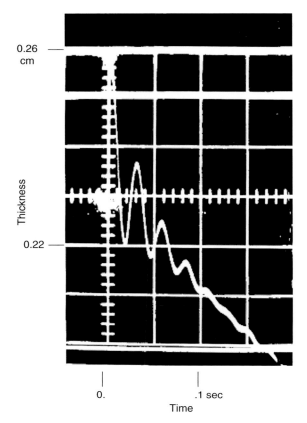

Figure 1.8. Oscilloscope trace showing thickness versus time for squeezing flow of a silicone polymer with a constant force. Reprinted from Lee et al., *J. Non-Newtonian Fluid Mech.*, **14**, 301 (1984).

η_0. At intermediate shear rates the viscosity decreases with increasing shear rate, a property known as *shear thinning*. It is believed that the viscosity approaches another limiting value at high shear rates, and this behavior is observed in polymer solutions, but it is extremely difficult to reach this regime experimentally with polymer melts. The zero-shear viscosity of linear polymers increases linearly with molecular weight until there are sufficient entanglements per chain, after which η_0 varies with the molecular weight to a power that is usually close to 3.4.

1.3.2 Melt Elasticity

The entangled polymer chains require time to move relative to one another following imposition of a stress. Hence, we can expect to find a characteristic time scale λ for the polymer melt such that for times $t \ll \lambda$, before the chains have had an opportunity to move within the entangled network, the system will respond like a rubbery solid, with a characteristic elastic modulus G. For times $t \gg \lambda$ the network has an opportunity to respond and the behavior is fluidlike. Figure 1.8 shows a particularly dramatic manifestation of the elastic character of a polymer melt, in this case a silicone polymer (i.e., a polymer in which the backbone consists of

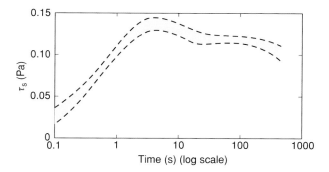

Figure 1.9. Shear stress following application of a constant shear rate of 1 s^{-1}. Data for eight samples of a low-density polyethylene fall within the shaded region. Reprinted with permission from Meissner, *Pure Appl. Chem.*, **42**, 553 (1975).

Si atoms in place of the carbons). This is a compression molding-like squeezing experiment carried out between two circular plates under constant force, with the spacing between the plates monitored as a function of time. We see a rapid change in the thickness, accompanied by damped oscillations, followed by a more gradual closing of the plates. An undamped oscillation would be a purely elastic response, which is energy conserving (sum of kinetic and potential energy = constant), while the gradual decay is a dissipative response typical of a viscous liquid. The elastic modulus required to represent this behavior is 6.65×10^5 pascal; the modulus for typical polyethylenes tends to be slightly higher. The characteristic response time for the network from this experiment seems to be on the order of 0.2 s; response times for carbon-based polymers range from this magnitude to several seconds and more, with the latter value typical of polyethylenes at processing temperatures. Figure 1.9, for example, shows the stress response in a linear low-density polyethylene following the imposition of a constant shear rate. The transient here is on the order of seconds.

The characteristic time constant for the polymer network response means that any dimensional analysis of a melt process will involve at least one more dimensionless group than would exist for a Newtonian fluid under the same conditions. If the process itself has a characteristic time, which we will denote t_p – the residence time on a spinline, for example – then a group that arises naturally is the ratio λ/t_p, the ratio of the characteristic time of the fluid to the characteristic time of the process. If this dimensionless group is large, the process occurs too quickly for the entangled network to adjust and the response is dominated by the network elasticity. If the group is small, however, the network can adjust and the response is dominated by viscous dissipative processes. This dimensionless ratio of time scales has come to be known as the *Deborah number*, denoted De, after the Biblical prophetess Deborah (from the Song of Deborah, Judges 5:5, "The mountains quaked [sometimes translated as "flowed"] at the presence of the Lord," suggesting that even solid mountains flow like a liquid on an appropriately long process time scale), and we speak of "high Deborah number processes" in which an elastic response dominates and "low Deborah number processes" in which a viscous response dominates.

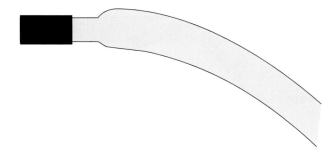

Figure 1.10. Extrudate swell, 5% solution of polyisobutylene in decalin. This drawing is based on the original photograph. Reprinted with permission from Metzner et al., *Chem. Eng. Progress*, **62** (12), 81 (1966). Copyright American Institute of Chemical Engineers.

1.3.3 Extrudate Swell

Extrudate swell is a dramatic manifestation of melt elasticity and is one of the distinguishing characteristics of polymer melts. When a low molar-mass liquid is extruded at a low Reynolds number from a cylindrical tube, the emerging jet has a diameter that is approximately 13% greater than the diameter of the tube because of the velocity profile rearrangement at the exit. (The high Reynolds number jet, in which the jet diameter is approximately 82% of the tube diameter, is the more familiar phenomenon. There is a smooth transition from slight swelling to contraction in the Reynolds number range from 1 to 50.) A molten polymer can swell to a diameter that is as much as 200% of the tube diameter. An example is shown in Figure 1.10. This phenomenon is caused by the relaxation of *normal stresses*, which are another manifestation of melt elasticity. Normal stresses are stresses transverse to the direction of shear; they are the fluid equivalent of the *Poynting effect* in nonlinear elasticity, in which torsion of an elastic rod causes a stress in the axial direction. Qualitatively, these stresses push out against the walls of the tube; when the polymer emerges and the walls are "removed," the polymer melt is able to expand. Large extrudate swell is obviously a concern in die design since the desired size is that of the extrudate, not the die. It is a particularly interesting issue in dies of noncircular shape, where the swell is likely to be uneven around the periphery, and the die shape may need to be different from that of the desired extrudate. Extrudate swell is also an issue in the design of the gate in an injection mold.

1.4 Polymer Chain Characteristics

There are several properties of the chain that are important in processing, in large measure because they affect the rheological properties or the solidification process.

1.4.1 Branching

We tend to visualize a polymer chain as a straight sequence, and some polymers exist only in this form. The synthetic chemistry can lead to other structures, however, either by design or because of competing reactions. One such structure is

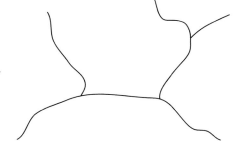

Figure 1.11. Schematic of branching.

branching, which is illustrated in Figure 1.11 for polyethylene. The linear molecule packs more easily in the solid state and hence is more crystalline and has a higher solid-phase density than the branched molecule (960–970 vs. 915–930 kg/m^3); the linear molecule is known as *high-density polyethylene* (HDPE), while the molecule with long branches is known as *low-density polyethylene* (LDPE). There is a synthetic route involving copolymerization of ethylene with 1-butene, 1-hexene, or 1-octene that produces a low-density molecule (density 915–940 kg/m^3) with short branches, known as *linear low-density polyethylene* (LLDPE). Branched polymers tend to be more shear thinning than linear molecules. The effect of branching on the zero-shear viscosity is complex because long branches change the nature of the network disentanglement process. Linear molecules disentangle by a process known as *reptation* ("snakelike motion"), in which the molecule moves along its backbone direction because it is precluded from large sideward motions by the presence of the neighboring chains in the network. This mechanism is hindered by the presence of long branches, making the disentanglement process more difficult. Hence, the zero-shear viscosity of a branched molecule is lower than that of a linear molecule of the same molecular weight for relatively short branches, but for sufficiently long branches the reverse occurs, and the zero-shear viscosity for a branched molecule can have a much stronger dependence on molecular weight than the 3.4-power characteristic of linear polymers.

1.4.2 Molecular Weight Distribution

The synthetic processes for manufacturing polymers result in a distribution of molecular weights within any batch. Two different moments of the distribution are commonly used to characterize a polymer. The mean of the distribution is known as the number-average molecular weight and denoted M_n. Let M_o be the molecular weight of the repeat unit, and let f_i be the fraction of the molecules containing i repeat units in the chain. The number-average molecular weight is then simply

$$M_n = M_o \sum_{i=1}^{\infty} i f_i \sim M_o \int_1^{\infty} i f(i) di. \qquad (1.1)$$

The integral approximation is useful if the distribution is represented by a continuous function. A second moment that is used is known as the *weight-average molecular weight*, denoted M_w and defined as

$$M_w = \frac{M_o \sum_{i=1}^{\infty} i^2 f_i}{\sum_{i=1}^{\infty} i f_i} \sim \frac{M_o \int_1^{\infty} i^2 f(i) di}{\int_1^{\infty} i f(i) di}. \tag{1.2}$$

The weight-average molecular weight is related to the variance of the distribution. The breadth of the distribution can be described in a number of ways using these two moments. The most common way to describe the breadth is by the *polydispersity*, defined as M_w/M_n, which is always greater than or equal to unity. A monodisperse polymer, in which all chains are exactly the same length, will have a polydispersity of 1. Nearly monodisperse polymers can be synthesized by special methods, but commercial polymers generally have very broad distributions. Condensation polymers (polyesters, polyamides, and polyacetals) have values of M_w/M_n close to the value of 2 that is predicted theoretically for the given chemistry. Linear polyethylenes typically have M_w/M_n values of about 4, while values for branched polyethylenes are typically 10 or greater. The transition from the zero-shear viscosity to shear thinning is very sharp for narrow-distribution polymers, while it is very gradual for polymers with a broad molecular weight distribution; this difference can be seen for the two sets of polystyrene data in Figure 1.7. Other rheological properties are similarly affected. The molecular weight distribution is typically obtained by chromatographic methods, but there are many difficulties in this approach, including the limited solubility of many polymers and the absence of reliable standards. Because of the strong dependence of rheological behavior on the molecular weight distribution, there is considerable activity in the use of rheological measurements as a means of obtaining online information about the distribution. The weight-average molecular weight is the appropriate value to use when applying the 3.4-power rule for the molecular-weight dependence of the zero-shear viscosity.

1.4.3 Transitions

Solid polymers undergo a variety of physical transitions. The most obvious from the processing perspective are crystallization and melting and the glass transition. These transitions are usually determined by differential scanning calorimetry (DSC), a thermal method in which the temperature of a very small sample is slowly changed and the heat flow is measured. (The glass transition temperature is also determined by mechanical rheometry, which we discuss in Chapter 9.) A transition is marked by a large change in the heat flux, characterizing the phase transition enthalpy. A typical DSC result is shown in Figure 1.12 for blends of branched and linear polyethylene. The broad melting peaks in the pure samples indicate that melting takes place over a finite temperature range, in contrast to the sharp melting temperatures of crystals of low molar-mass materials. The presence of two peaks in the blends indicates the formation of separate crystalline phases of the two components, while the formation of a single peak at low LDPE concentrations indicates miscibility and the

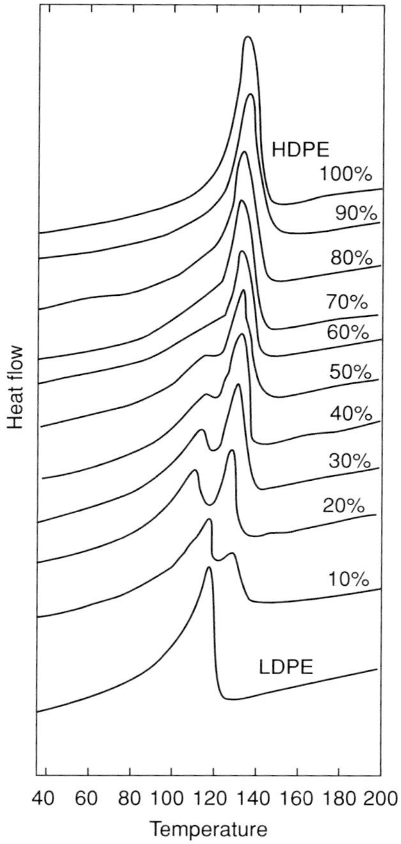

Figure 1.12. Differential scanning calorimetry traces for blends of HDPE and LDPE. Reprinted from Lee and Denn, *Polym. Eng. Sci.*, **40**, 1132 (2000).

formation of a single crystalline phase. The glass transition and melting temperatures of some common polymers are shown in Table 1.1. The melting temperatures are the peak temperatures from the DSC trace.

BIBLIOGRAPHICAL NOTES

The second edition of the *Encyclopedia of Polymer Science and Engineering*, published in seventeen volumes by John Wiley between 1985 and 1989, and the third edition, published in twelve volumes between 2003 and 2004, are excellent sources for all of the topics addressed in this introductory chapter. Most are also addressed at least superficially in any introductory textbook on polymers. A particularly readable introduction to the physics of polymers can be found in

Grossberg, A. Yu., and A. R. Khokhlov, *Giant Molecules*, Academic Press, San Diego, 1997.

Detailed treatments are in

Graessley, W. W., *Polymeric Liquids and Networks: Structure and Properties*, Garland Science, New York, 2004.
Rubinstein, M., and R. H. Colby, *Polymer Physics*, Oxford University Press, New York, 2003.

2 Fundamentals

2.1 Introduction

Polymer flow in any melt processing geometry is governed by three fundamental principles of physics: conservation of mass, conservation of linear momentum, and conservation of energy. Linear momentum is a vector quantity that must be conserved in each of three independent coordinate directions, so we must expect five conservation statements in the most general case. The natural language of these conservation statements is differential and integral calculus (recall that Newton invented the calculus to enable him to describe problems of motion); in particular, because we have four independent variables – time and three spatial variables – our language will employ partial derivatives, and the conservation equations will be stated as partial differential equations. The problems we will address in this text do *not* generally require familiarity with methods of solution of partial differential equations because the equations will usually simplify to forms that can be analyzed using elementary concepts of the calculus of one independent variable. (An apt linguistic analogy might be the contrast between understanding basic prose – our task here – and writing poetry.) Hence, the subject is open to any student who has completed a basic sequence in calculus.

The conservation equations are derived in many textbooks on fluid mechanics and transport phenomena, and we shall simply state them here, with an explanation of the meanings of terms where appropriate.

2.2 Conservation Principles

2.2.1 Conservation of Mass

The principle of mass conservation simply states that mass is neither created nor destroyed. (Note that individual species *can* be created and destroyed.) Thus, in any arbitrarily defined region of space, known as a *control volume*, the rate of change of mass contained within the volume must equal the net rate at which mass crosses the boundaries. To express mass and flow we clearly require the density, ρ, and the velocity, $\mathbf{v}$; $\mathbf{v}$ is a vector with components v_x, v_y, and v_z in the three Cartesian coordinate directions.

Conservation of mass is expressed in the following form, known as the *continuity equation*:

$$\frac{\partial \rho}{\partial t} = -\frac{\partial}{\partial x}\rho v_x - \frac{\partial}{\partial y}\rho v_y - \frac{\partial}{\partial z}\rho v_z. \tag{2.1}$$

That is, the rate of change of the density with time at any position in space ($\partial \rho/\partial t$) equals the negative sum of three terms, each of which is the spatial rate of change of a mass flux, or a mass flow rate on a unit area basis in a specific direction. An alternative and completely equivalent form is obtained by applying the product rule to each term on the right side of Equation 2.1:

$$\frac{D\rho}{Dt} = \frac{\partial \rho}{\partial t} + v_x\frac{\partial \rho}{\partial x} + v_y\frac{\partial \rho}{\partial y} + v_z\frac{\partial \rho}{\partial z} = -\rho\left(\frac{\partial v_x}{\partial x} + \frac{\partial v_y}{\partial y} + \frac{\partial v_z}{\partial z}\right). \tag{2.2}$$

The symbol $D\rho/Dt$, known as the *substantial derivative*, is shorthand for the sum of the four terms on the left side of Equation 2.2.

Throughout this text we will assume that polymer melts are incompressible liquids, by which we mean that the density never changes with position or time. This is clearly an approximation that must be relaxed in some applications – injection molding, for example, where the compressibility of the melt becomes important because of the extremely high pressures – but the incompressibility assumption will suffice for our purpose here. If the density never changes in time or space, rates of change with respect to these variables (i.e., derivatives) must be zero ($\partial \rho/\partial t = 0$, $\partial \rho/\partial x = 0$, etc.), and the continuity equation simplifies to

$$\nabla \cdot \mathbf{v} \equiv \frac{\partial v_x}{\partial x} + \frac{\partial v_y}{\partial y} + \frac{\partial v_z}{\partial z} = 0. \tag{2.3a}$$

The symbol $\nabla \cdot \mathbf{v}$, known as the *divergence* of the velocity vector $\mathbf{v}$, is shorthand for the three terms on the left of Equation 2.3a. We can, of course, choose to employ any coordinate system we wish, and for many problems – flow in a round pipe, for example – rectangular Cartesian coordinates are inconvenient. It is straightforward to transform Equation 2.3a to other coordinates; in a cylindrical (r, θ, z) coordinate system, for example, the equivalent form of Equation 2.3a is

$$\nabla \cdot \mathbf{v} \equiv \frac{1}{r}\frac{\partial}{\partial r}(rv_r) + \frac{1}{r}\frac{\partial v_\theta}{\partial \theta} + \frac{\partial v_z}{\partial z} = 0. \tag{2.3b}$$

Here, v_r, v_θ, and v_z are the r (radial), θ (circumferential), and z (axial) components of the velocity at each position. The equation $\nabla \cdot \mathbf{v} = 0$ is shown in Table 2.1 for the three most common coordinate systems: rectangular Cartesian, cylindrical, and spherical.

Let us see what we can learn from the continuity equation. Suppose we have flow between two converging planes (Figure 2.1), such as we might experience in a film die. The origin of a cylindrical (r, θ, z) coordinate system is put at the point where the planes would meet, and the angle θ is measured from the plane of symmetry. z is the "neutral" direction into the paper, and we assume there is no flow

Table 2.1. *Continuity equation for an incompressible fluid*

$$\nabla \cdot \mathbf{v} = 0$$

Rectangular Cartesian (x, y, z) coordinates:

$$\frac{\partial v_x}{\partial x} + \frac{\partial v_y}{\partial y} + \frac{\partial v_z}{\partial z} = 0$$

Cylindrical (r, θ, z) coordinates:

$$\frac{1}{r}\frac{\partial}{\partial r}(rv_r) + \frac{1}{r}\frac{\partial v_\theta}{\partial \theta} + \frac{\partial v_z}{\partial z} = 0$$

Spherical (r, θ, ϕ) coordinates:

$$\frac{1}{r^2}\frac{\partial}{\partial r}(r^2 v_r) + \frac{1}{r\sin\theta}\frac{\partial}{\partial \theta}(v_\theta \sin\theta) + \frac{1}{r\sin\theta}\frac{\partial v_\phi}{\partial \phi} = 0$$

in the z direction. Away from the entry and exit of the converging region it seems reasonable to suppose that elements of the fluid move along radial lines; that is, if we were to put a spot of dye into the fluid at a distance r from the origin and an angle θ from the midplane, at a later time we would find the spot of dye at a new value of r, closer to the origin, but at the same angle θ from the midplane. This presumption is equivalent to the mathematical statement that $v_\theta = 0$ everywhere, since the fluid containing the dye would move in the azimuthal direction to a new value of θ if there were an azimuthal velocity. Now, if v_z and v_θ are both zero, only the first term remains in Equation 2.3b, and we may write

$$\frac{1}{r}\frac{\partial}{\partial r}(rv_r) = 0. \tag{2.4}$$

Since the rate of change of (rv_r) with respect to r is zero, it therefore follows that rv_r is independent of r. rv_r may still depend on θ, however, so we write

$$rv_r = \phi(\theta), \tag{2.5a}$$

where $\phi(\theta)$ is a function of θ whose form we do not know; equivalently,

$$v_r = \frac{\phi(\theta)}{r}. \tag{2.5b}$$

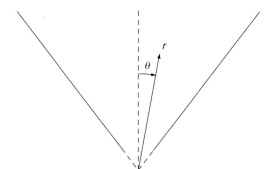

Figure 2.1. Converging flow between "infinite" parallel planes.

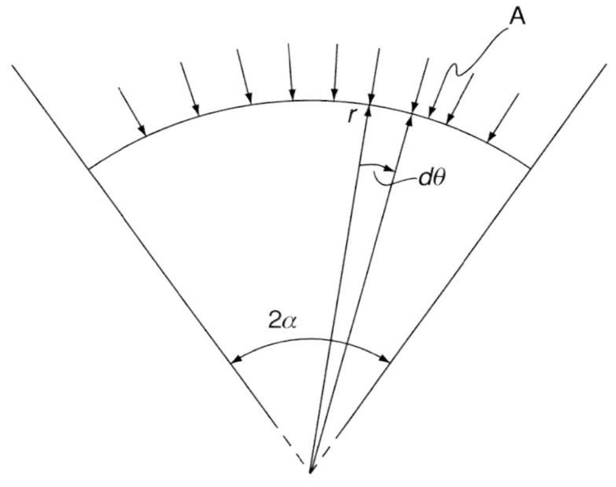

Figure 2.2. Flow across a surface at a fixed radial position in radial flow between infinite converging planes.

Thus, the continuity equation tells us that the radial velocity at any angle varies inversely with the radius.[*]

2.2.2 Conservation of Linear Momentum

Newton's second principle states that the rate of change of momentum of a particle equals the sum of the imposed forces. For a fluid, this principle is stated as follows for each direction in a rectangular Cartesian coordinate system:

$$x: \rho \left(\frac{\partial v_x}{\partial t} + v_x \frac{\partial v_x}{\partial x} + v_y \frac{\partial v_x}{\partial y} + v_z \frac{\partial v_x}{\partial z} \right) = -\frac{\partial p}{\partial x} + \frac{\partial \tau_{xx}}{\partial x} + \frac{\partial \tau_{xy}}{\partial y} + \frac{\partial \tau_{xz}}{\partial z} + \rho g_x, \quad (2.6a)$$

$$y: \rho \left(\frac{\partial v_y}{\partial t} + v_x \frac{\partial v_y}{\partial x} + v_y \frac{\partial v_y}{\partial y} + v_z \frac{\partial v_y}{\partial z} \right) = -\frac{\partial p}{\partial y} + \frac{\partial \tau_{yx}}{\partial x} + \frac{\partial \tau_{yy}}{\partial y} + \frac{\partial \tau_{yz}}{\partial z} + \rho g_y, \quad (2.6b)$$

$$z: \rho \left(\frac{\partial v_z}{\partial t} + v_x \frac{\partial v_z}{\partial x} + v_y \frac{\partial v_z}{\partial y} + v_z \frac{\partial v_z}{\partial z} \right) = -\frac{\partial p}{\partial z} + \frac{\partial \tau_{zx}}{\partial x} + \frac{\partial \tau_{zy}}{\partial y} + \frac{\partial \tau_{zz}}{\partial z} + \rho g_z. \quad (2.6c)$$

Equations 2.6a–c are known collectively as the *Cauchy momentum equation*. Before we discuss the meanings of the terms in the equations, it is useful to note that the y-direction equation can be obtained from the x-direction equation simply by permuting indices: $x, y, z \rightarrow y, z, x$. Similarly, the z direction is obtained from $y, z, x \rightarrow z, x, y$.

The terms on the left side of the Cauchy momentum equation sum to the rate of change of momentum, or inertia, on a unit volume basis. There are four terms because momentum in a given direction changes as the velocity changes with time and as a fluid element changes direction. The first term on the right is the rate of

[*] The inverse radial dependence is expected. Consider the surface denoted "A" in Figure 2.2, which is the arc of a circle of radius r. Denote the mass flow rate across that surface as ρQ, where Q is the volumetric flow rate. (Volumetric flow rate = volume/time, while density = mass/volume, so ρQ has dimensions of mass/volume × volume/time = mass/time.) $Q = \overline{v_r}(2\alpha r)L$, where 2α is the angle between the planes, L is the length into the page, and $\overline{v_r}$ is the average radial velocity (flow rate = average velocity × surface area). Thus, $\overline{v_r} = Q/2\alpha L r$. $Q/2\alpha L$ is simply a constant (the flow rate is the same across any arc, since the fluid is incompressible), so we easily establish the inverse radial dependence for the average velocity.

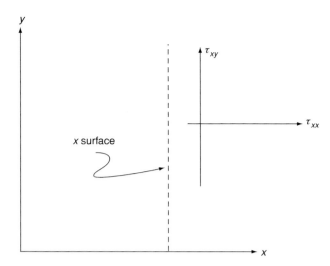

Figure 2.3. Schematic of stresses acting on an x surface.

change of pressure with position, or the pressure gradient. It is helpful to recall that a derivative is the limit of a ratio of differences ($\partial p/\partial x \approx$ small change in pressure/small distance in the x direction, for example). This term then reflects a force difference across a material element (pressure $\times$ area) per unit volume (area $\times$ distance). The final term in each equation is the gravitational force per unit volume acting on the fluid element; g_x, for example, is the component of the gravitational acceleration in the x direction.

The nine terms $\tau_{xx}, \tau_{xy}, \ldots, \tau_{zz}$ are known as the components of the *extra stress*. These terms reflect the internal forces that fluid elements exert on one another because of the deformation. The first subscript denotes the surface in the fluid that is experiencing the force, while the second subscript denotes the component direction. For example (Figure 2.3), τ_{xx} is a *normal stress* because it is a stress on an "x surface" acting in the x direction; τ_{xy} is a *shear stress* because it acts parallel to the x surface. These stresses depend on the deformation, and they play a central role in the analysis of the polymer processes, for it is through the stresses that differences between materials manifest themselves.

The principle of the conservation of angular momentum is often used to argue for the symmetry of the extra stress; that is, $\tau_{xy} = \tau_{yx}$, and so forth. In that case there are six independent components, not nine. The angular momentum argument requires an explicit but often unstated assumption that there is no structure in the fluid that is capable of generating local torques, which seems generally to be the case for polymers (except perhaps for liquid crystalline polymers), and, except for a brief introduction to liquid crystals in Chapter 13, we will assume stress symmetry throughout.

There is one aspect of the momentum equation that often causes confusion. The pressure is an isotropic stress; that is, it acts equally in all coordinate directions. Thus, the *total* normal stress acting on, say, the x surface is the sum of the pressure and the extra stress τ_{xx}. We have implicitly adopted a sign convention that says that pressures are positive when putting a fluid element in compression, but extra stresses are positive when putting an element in tension; this is the reason that p and τ_{xx} enter Equation 2.6a with opposite signs.[*] Thus, the *total* stress acting on an x surface, which we denote σ_{xx}, is

$$\sigma_{xx} = -p + \tau_{xx}. \tag{2.7a}$$

Similarly,

$$\sigma_{yy} = -p + \tau_{yy}, \tag{2.7b}$$

$$\sigma_{zz} = -p + \tau_{zz}. \tag{2.7c}$$

The Cauchy momentum equation can be transformed to other coordinate systems that might be more useful for particular problems. Flow in a capillary, for example, will be described most naturally in cylindrical (r, θ, z) coordinates. The momentum equation is shown in Table 2.2 in the three commonly used coordinate systems. The equation is often written in the shorthand vector form

$$\rho \frac{D\mathbf{v}}{Dt} = -\nabla p + \nabla \cdot \boldsymbol{\tau} + \rho \mathbf{g}, \tag{2.8}$$

where the meanings of the symbols can be discerned by comparison with the component equations.

Finally, consider the sum

$$\mathcal{P} = p + \rho g h, \tag{2.9}$$

where h is the height above an arbitrary datum. It can be shown from trigonometric arguments that

$$g_x = -g \frac{\partial h}{\partial x}, \quad g_y = -g \frac{\partial h}{\partial y}, \quad \text{and} \quad g_z = -g \frac{\partial h}{\partial z}.$$

Thus, an alternative form of Equation 2.8 for incompressible liquids $(\rho = \text{constant})$ is

$$\rho \frac{D\mathbf{v}}{Dt} = -\nabla \mathcal{P} + \nabla \cdot \tau; \tag{2.10}$$

that is, we simply replace p by $\mathcal{P}$ in any of the equations in Table 2.2 and drop the gravity term. This form of the equation is usually the most convenient for confined flows, such as flow in an extrusion die, but not for flows with free surfaces, such as fiber spinning. $\mathcal{P}$ is sometimes known as the *equivalent pressure*.[**]

---

[*] The opposite convention is also used by some authors for the stress, in which case Equation 2.7a would be written $\sigma_{xx} = p + \tau_{xx}$. This can be confusing to the student new to the subject, but the convention clearly must be irrelevant to the final physical result. Readers who have studied thermodynamics will recall a similar situation with regard to the sign convention for work; work is sometimes defined as positive if done *on* the system, sometimes as positive if done *by* the system.

[**] Consider a case in which there is no flow, so $v_x = v_y = v_z = 0$. Also, all $\tau_{xx} = \tau_{xy} = \ldots = 0$ since there is no deformation that can generate stresses. The three components of the momentum equation then reduce to $\partial \mathcal{P}/\partial x = 0$, $\partial \mathcal{P}/\partial y = 0$, $\partial \mathcal{P}/\partial z = 0$; that is, $\mathcal{P}$ is independent of x, y, and z and hence is a constant in space. Suppose gravity points in the negative z direction; we then have $\mathcal{P} = p + \rho g z = \text{constant}$ with $\mathcal{P}$ independent of x and y at each value of z. This is simply the basic equation of hydrostatics, which describes, for example, the operation of a manometer in which differences in liquid height are used to measure pressure differences.

Table 2.2. *Cauchy momentum equation*

$$\rho \frac{D\mathbf{v}}{Dt} = -\nabla p + \nabla \cdot \tau + \rho \mathbf{g}$$

Rectangular Cartesian (x, y, z) coordinates:

x component: $\rho \left(\dfrac{\partial v_x}{\partial t} + v_x \dfrac{\partial v_x}{\partial x} + v_y \dfrac{\partial v_x}{\partial y} + v_z \dfrac{\partial v_x}{\partial z} \right) = -\dfrac{\partial p}{\partial x} + \dfrac{\partial \tau_{xx}}{\partial x} + \dfrac{\partial \tau_{xy}}{\partial y} + \dfrac{\partial \tau_{xz}}{\partial z} + \rho g_x$

y component: $\rho \left(\dfrac{\partial v_y}{\partial t} + v_x \dfrac{\partial v_y}{\partial x} + v_y \dfrac{\partial v_y}{\partial y} + v_z \dfrac{\partial v_y}{\partial z} \right) = -\dfrac{\partial p}{\partial y} + \dfrac{\partial \tau_{yx}}{\partial x} + \dfrac{\partial \tau_{yy}}{\partial y} + \dfrac{\partial \tau_{yz}}{\partial z} + \rho g_y$

z component: $\rho \left(\dfrac{\partial v_z}{\partial t} + v_x \dfrac{\partial v_z}{\partial x} + v_y \dfrac{\partial v_z}{\partial y} + v_z \dfrac{\partial v_z}{\partial z} \right) = -\dfrac{\partial p}{\partial z} + \dfrac{\partial \tau_{zx}}{\partial x} + \dfrac{\partial \tau_{zy}}{\partial y} + \dfrac{\partial \tau_{zz}}{\partial z} + \rho g_z$

Cylindrical (r, θ, z) coordinates:

r component: $\rho \left(\dfrac{\partial v_r}{\partial t} + v_r \dfrac{\partial v_r}{\partial r} + \dfrac{v_\theta}{r} \dfrac{\partial v_r}{\partial \theta} - \dfrac{v_\theta^2}{r} + v_z \dfrac{\partial v_r}{\partial z} \right) = -\dfrac{\partial p}{\partial r}$

$$+ \frac{1}{r} \frac{\partial}{\partial r}(r \tau_{rr}) + \frac{1}{r} \frac{\partial \tau_{r\theta}}{\partial \theta} - \frac{\tau_{\theta\theta}}{r} + \frac{\partial \tau_{rz}}{\partial z} + \rho g_r$$

θ component: $\rho \left(\dfrac{\partial v_\theta}{\partial t} + v_r \dfrac{\partial v_\theta}{\partial r} + \dfrac{v_\theta}{r} \dfrac{\partial v_\theta}{\partial \theta} + \dfrac{v_r v_\theta}{r} + v_z \dfrac{\partial v_\theta}{\partial z} \right) = -\dfrac{1}{r} \dfrac{\partial p}{\partial \theta}$

$$+ \frac{1}{r^2} \frac{\partial}{\partial r}(r^2 \tau_{r\theta}) + \frac{1}{r} \frac{\partial \tau_{\theta\theta}}{\partial \theta} + \frac{\partial \tau_{\theta z}}{\partial z} + \rho g_\theta$$

z component: $\rho \left(\dfrac{\partial v_z}{\partial t} + v_r \dfrac{\partial v_z}{\partial r} + \dfrac{v_\theta}{r} \dfrac{\partial v_z}{\partial \theta} + v_z \dfrac{\partial v_z}{\partial z} \right) = -\dfrac{\partial p}{\partial z}$

$$+ \frac{1}{r} \frac{\partial}{\partial r}(r \tau_{rz}) + \frac{1}{r} \frac{\partial \tau_{\theta z}}{\partial \theta} + \frac{\partial \tau_{zz}}{\partial z} + \rho g_z$$

Spherical (r, θ, ϕ) coordinates:

r component: $\rho \left(\dfrac{\partial v_r}{\partial t} + v_r \dfrac{\partial v_r}{\partial r} + \dfrac{v_\theta}{r} \dfrac{\partial v_r}{\partial \theta} + \dfrac{v_\phi}{r \sin \theta} \dfrac{\partial v_r}{\partial \phi} - \dfrac{v_\theta^2 + v_\phi^2}{r} \right)$

$$= -\frac{\partial p}{\partial r} + \frac{1}{r^2} \frac{\partial}{\partial r}(r^2 \tau_{rr}) + \frac{1}{r \sin \theta} \frac{\partial}{\partial \theta}(\tau_{r\theta} \sin \theta) + \frac{1}{r \sin \theta} \frac{\partial \tau_{r\phi}}{\partial \phi} - \frac{\tau_{\theta\theta} + \tau_{\phi\phi}}{r} + \rho g_r$$

θ component: $\rho \left(\dfrac{\partial v_\theta}{\partial t} + v_r \dfrac{\partial v_\theta}{\partial r} + \dfrac{v_\theta}{r} \dfrac{\partial v_\theta}{\partial \theta} + \dfrac{v_\phi}{r \sin \theta} \dfrac{\partial v_\theta}{\partial \phi} + \dfrac{v_r v_\theta}{r} - \dfrac{v_\phi^2 \cot \theta}{r} \right)$

$$= -\frac{1}{r} \frac{\partial p}{\partial \theta} + \frac{1}{r^2} \frac{\partial}{\partial r}(r^2 \tau_{r\theta}) + \frac{1}{r \sin \theta} \frac{\partial}{\partial \theta}(\tau_{\theta\theta} \sin \theta) + \frac{1}{r \sin \theta} \frac{\partial \tau_{\theta\phi}}{\partial \phi} + \frac{\tau_{r\theta}}{r} - \frac{\cot \theta}{r} \tau_{\phi\phi} + \rho g_\theta$$

ϕ component: $\rho \left(\dfrac{\partial v_\phi}{\partial t} + v_r \dfrac{\partial v_\phi}{\partial r} + \dfrac{v_\theta}{r} \dfrac{\partial v_\phi}{\partial \theta} + \dfrac{v_\phi}{r \sin \theta} \dfrac{\partial v_\phi}{\partial \phi} + \dfrac{v_\phi v_r}{r} + \dfrac{v_\theta v_\phi}{r} \cot \theta \right)$

$$= -\frac{1}{r \sin \theta} \frac{\partial p}{\partial \phi} + \frac{1}{r^2} \frac{\partial}{\partial r}(r^2 \tau_{r\phi}) + \frac{1}{r} \frac{\partial \tau_{\theta\phi}}{\partial \theta} + \frac{1}{r \sin \theta} \frac{\partial \tau_{\phi\phi}}{\partial \phi} + \frac{\tau_{r\phi}}{r} + \frac{2 \cot \theta}{r} \tau_{\theta\phi} + \rho g_\phi$$

2.2.3 Newtonian Fluid

We need to digress briefly from the logical presentation of conservation equations to introduce a particular *constitutive equation*, that is, an equation relating the extra stress to the deformation of the fluid. For low molecular weight liquids, such as water and glycerol, it is found experimentally that the stress to deform the liquid is directly proportional to the instantaneous rate of deformation. The rigorous mathematical expression of this experimental observation for an incompressible liquid,

Table 2.3. *Incompressible Newtonian fluid*

Rectangular Cartesian (x, y, z) coordinates:

$$\tau_{xx} = 2\eta \frac{\partial v_x}{\partial x} \qquad \tau_{yy} = 2\eta \frac{\partial v_y}{\partial y} \qquad \tau_{zz} = 2\eta \frac{\partial v_z}{\partial z}$$

$$\tau_{xy} = \tau_{yx} = \eta \left(\frac{\partial v_x}{\partial y} + \frac{\partial v_y}{\partial x} \right) \qquad \tau_{xz} = \tau_{zx} = \eta \left(\frac{\partial v_x}{\partial z} + \frac{\partial v_z}{\partial x} \right)$$

$$\tau_{yz} = \tau_{zy} = \eta \left(\frac{\partial v_y}{\partial z} + \frac{\partial v_z}{\partial y} \right)$$

Cylindrical (r, θ, z) coordinates:

$$\tau_{rr} = 2\eta \frac{\partial v_r}{\partial r} \qquad \tau_{\theta\theta} = 2\eta \left(\frac{1}{r} \frac{\partial v_\theta}{\partial \theta} + \frac{v_r}{r} \right) \qquad \tau_{zz} = 2\eta \frac{\partial v_z}{\partial z}$$

$$\tau_{r\theta} = \tau_{\theta r} = \eta \left[r \frac{\partial}{\partial r} \left(\frac{v_\theta}{r} \right) + \frac{1}{r} \frac{\partial v_r}{\partial \theta} \right] \qquad \tau_{z\theta} = \tau_{\theta z} = \eta \left(\frac{\partial v_\theta}{\partial z} + \frac{1}{r} \frac{\partial v_z}{\partial \theta} \right)$$

$$\tau_{rz} = \tau_{zr} = \eta \left(\frac{\partial v_z}{\partial r} + \frac{\partial v_r}{\partial z} \right)$$

Spherical (r, θ, ϕ) coordinates:

$$\tau_{rr} = 2\eta \frac{\partial v_r}{\partial r} \qquad \tau_{\theta\theta} = 2\eta \left(\frac{1}{r} \frac{\partial v_\theta}{\partial \theta} + \frac{v_r}{r} \right) \qquad \tau_{\phi\phi} = 2\eta \left(\frac{1}{r \sin\theta} \frac{\partial v_\phi}{\partial \phi} + \frac{v_r}{r} + \frac{v_\theta \cot\theta}{r} \right)$$

$$\tau_{r\theta} = \tau_{\theta r} = \eta \left[r \frac{\partial}{\partial r} \left(\frac{v_\theta}{r} \right) + \frac{1}{r} \frac{\partial v_r}{\partial \theta} \right] \qquad \tau_{\theta\phi} = \tau_{\phi\theta} = \eta \left[\frac{\sin\theta}{r} \frac{\partial}{\partial \theta} \left(\frac{v_\phi}{\sin\theta} \right) + \frac{1}{r \sin\theta} \frac{\partial v_\theta}{\partial \phi} \right]$$

$$\tau_{\phi r} = \tau_{r\phi} = \eta \left[\frac{1}{r \sin\theta} \frac{\partial v_r}{\partial \phi} + r \frac{\partial}{\partial r} \left(\frac{v_\phi}{r} \right) \right]$$

which must be formulated to satisfy the principle of physics that a description of nature must be invariant to changes in the frame of reference of the observer, is

$$\tau_{xy} = \tau_{yx} = \eta \left(\frac{\partial v_x}{\partial y} + \frac{\partial v_y}{\partial x} \right), \tag{2.11a}$$

$$\tau_{xz} = \tau_{zx} = \eta \left(\frac{\partial v_x}{\partial z} + \frac{\partial v_z}{\partial x} \right), \tag{2.11b}$$

$$\tau_{yz} = \tau_{zy} = \eta \left(\frac{\partial v_y}{\partial z} + \frac{\partial v_z}{\partial y} \right), \tag{2.11c}$$

$$\tau_{xx} = 2\eta \frac{\partial v_x}{\partial x}, \quad \tau_{yy} = 2\eta \frac{\partial v_y}{\partial y}, \quad \tau_{zz} = 2\eta \frac{\partial v_z}{\partial z}. \tag{2.11d,e,f}$$

The single parameter, η, is known as the *viscosity*, and it is a measure of the resistance to deformation. η will depend in general on temperature and pressure. The equivalent equations for the extra stress in other coordinate systems are shown in Table 2.3.

When Equations 2.11a–f are substituted into the momentum equation, Equation 2.6, with the assumption that η is constant in space (that is, that temperature and pressure changes are sufficiently small that their effects on the viscosity can be ignored), we obtain, after some simplification, the *Navier-Stokes equations* for an incompressible Newtonian fluid:

$$x: \rho \left(\frac{\partial v_x}{\partial t} + v_x \frac{\partial v_x}{\partial x} + v_y \frac{\partial v_x}{\partial y} + v_z \frac{\partial v_x}{\partial z} \right) = -\frac{\partial \mathcal{P}}{\partial x} + \eta \left(\frac{\partial^2 v_x}{\partial x^2} + \frac{\partial^2 v_x}{\partial y^2} + \frac{\partial^2 v_x}{\partial z^2} \right),$$
(2.12a)

$$y: \rho \left(\frac{\partial v_y}{\partial t} + v_x \frac{\partial v_y}{\partial x} + v_y \frac{\partial v_y}{\partial y} + v_z \frac{\partial v_y}{\partial z} \right) = -\frac{\partial \mathcal{P}}{\partial y} + \eta \left(\frac{\partial^2 v_y}{\partial x^2} + \frac{\partial^2 v_y}{\partial y^2} + \frac{\partial^2 v_y}{\partial z^2} \right),$$
(2.12b)

$$z: \rho \left(\frac{\partial v_z}{\partial t} + v_x \frac{\partial v_z}{\partial x} + v_y \frac{\partial v_z}{\partial y} + v_z \frac{\partial v_z}{\partial z} \right) = -\frac{\partial \mathcal{P}}{\partial z} + \eta \left(\frac{\partial^2 v_z}{\partial x^2} + \frac{\partial^2 v_z}{\partial y^2} + \frac{\partial^2 v_z}{\partial z^2} \right), \quad (2.12c)$$

or, symbolically,

$$\rho \frac{D\mathbf{v}}{Dt} = -\nabla \mathcal{P} + \eta \nabla^2 \mathbf{v}.$$
(2.13)

The equivalent equations in other coordinate systems are shown in Table 2.4. These equations *do not* describe the motion of most polymeric liquids, but they do provide a useful frame of reference and are often adequate to describe the flow behavior of polyesters, nylons, and polycarbonates at low shear rates.

2.2.4 Creeping Flow

The terms on the left side of the Cauchy momentum equation, which we have written symbolically as $\rho(D\mathbf{v}/Dt)$, represent the contribution of inertial effects to the momentum balance. The inertial effects are negligible relative to the stresses generated within the fluid in most polymer processing operations, and to a very good approximation the inertial terms can usually be dropped. (Commercial fiber spinning is an exception.) For a Newtonian fluid the relative contribution of inertial and viscous terms is expressed as a dimensionless group known as the *Reynolds number*,

$$\mathrm{Re} \equiv \frac{Dv\rho}{\eta},$$
(2.14)

where D is a characteristic length and v a characteristic velocity (the exit diameter and velocity in an extrusion die, for example). We can show formally for problems with only one characteristic length that the inertial terms can be neglected whenever $\mathrm{Re} \ll 1$, which is common in polymer processing, in which case we obtain the *creeping flow equations:*

$$\mathbf{0} = -\nabla \mathcal{P} + \nabla \cdot \boldsymbol{\tau}.$$
(2.15)

For a Newtonian liquid the creeping flow equations are often called the *Stokes equations*:

$$\mathbf{0} = -\nabla \mathcal{P} + \eta \nabla^2 \mathbf{v}.$$
(2.16)

Table 2.4. *Navier-Stokes equations for an incompressible Newtonian fluid (in terms of* $\mathcal{P} = p + \rho g h$)

$$\rho \frac{D\mathbf{v}}{Dt} = -\nabla \mathcal{P} + \eta \nabla^2 \mathbf{v}$$

Rectangular Cartesian (*x*, *y*, *z*) coordinates:

x component: $\rho \left(\dfrac{\partial v_x}{\partial t} + v_x \dfrac{\partial v_x}{\partial x} + v_y \dfrac{\partial v_x}{\partial y} + v_z \dfrac{\partial v_x}{\partial z} \right) = -\dfrac{\partial \mathcal{P}}{\partial x} + \eta \left(\dfrac{\partial^2 v_x}{\partial x^2} + \dfrac{\partial^2 v_x}{\partial y^2} + \dfrac{\partial^2 v_x}{\partial z^2} \right)$

y component: $\rho \left(\dfrac{\partial v_y}{\partial t} + v_x \dfrac{\partial v_y}{\partial x} + v_y \dfrac{\partial v_y}{\partial y} + v_z \dfrac{\partial v_y}{\partial z} \right) = -\dfrac{\partial \mathcal{P}}{\partial y} + \eta \left(\dfrac{\partial^2 v_y}{\partial x^2} + \dfrac{\partial^2 v_y}{\partial y^2} + \dfrac{\partial^2 v_y}{\partial z^2} \right)$

z component: $\rho \left(\dfrac{\partial v_z}{\partial t} + v_x \dfrac{\partial v_z}{\partial x} + v_y \dfrac{\partial v_z}{\partial y} + v_z \dfrac{\partial v_z}{\partial z} \right) = -\dfrac{\partial \mathcal{P}}{\partial z} + \eta \left(\dfrac{\partial^2 v_z}{\partial x^2} + \dfrac{\partial^2 v_z}{\partial y^2} + \dfrac{\partial^2 v_z}{\partial z^2} \right)$

Cylindrical (*r*, *θ*, *z*) coordinates:

r component: $\rho \left(\dfrac{\partial v_r}{\partial t} + v_r \dfrac{\partial v_r}{\partial r} + \dfrac{v_\theta}{r} \dfrac{\partial v_r}{\partial \theta} - \dfrac{v_\theta^2}{r} + v_z \dfrac{\partial v_r}{\partial z} \right) = -\dfrac{\partial \mathcal{P}}{\partial r}$

$$+ \eta \left[\dfrac{\partial}{\partial r} \left(\dfrac{1}{r} \dfrac{\partial}{\partial r}(r v_r) \right) + \dfrac{1}{r^2} \dfrac{\partial^2 v_r}{\partial \theta^2} - \dfrac{2}{r^2} \dfrac{\partial v_\theta}{\partial \theta} + \dfrac{\partial^2 v_r}{\partial z^2} \right]$$

θ component: $\rho \left(\dfrac{\partial v_\theta}{\partial t} + v_r \dfrac{\partial v_\theta}{\partial r} + \dfrac{v_\theta}{r} \dfrac{\partial v_\theta}{\partial \theta} + \dfrac{v_r v_\theta}{r} + v_z \dfrac{\partial v_\theta}{\partial z} \right) = -\dfrac{1}{r} \dfrac{\partial \mathcal{P}}{\partial \theta}$

$$+ \eta \left[\dfrac{\partial}{\partial r} \left(\dfrac{1}{r} \dfrac{\partial}{\partial r}(r v_\theta) \right) + \dfrac{1}{r^2} \dfrac{\partial^2 v_\theta}{\partial \theta^2} + \dfrac{2}{r^2} \dfrac{\partial v_r}{\partial \theta} + \dfrac{\partial^2 v_\theta}{\partial z^2} \right]$$

z component: $\rho \left(\dfrac{\partial v_z}{\partial t} + v_r \dfrac{\partial v_z}{\partial r} + \dfrac{v_\theta}{r} \dfrac{\partial v_z}{\partial \theta} + v_z \dfrac{\partial v_z}{\partial z} \right) = -\dfrac{\partial \mathcal{P}}{\partial z}$

$$+ \eta \left[\dfrac{1}{r} \dfrac{\partial}{\partial r} \left(r \dfrac{\partial v_z}{\partial r} \right) + \dfrac{1}{r^2} \dfrac{\partial^2 v_z}{\partial \theta^2} + \dfrac{\partial^2 v_z}{\partial z^2} \right]$$

Spherical (*r*, *θ*, *φ*) coordinates:

r component: $\rho \left(\dfrac{\partial v_r}{\partial t} + v_r \dfrac{\partial v_r}{\partial r} + \dfrac{v_\theta}{r} \dfrac{\partial v_r}{\partial \theta} + \dfrac{v_\phi}{r \sin\theta} \dfrac{\partial v_r}{\partial \phi} - \dfrac{v_\theta^2 + v_\phi^2}{r} \right) = -\dfrac{\partial \mathcal{P}}{\partial r}$

$$+ \eta \left[\dfrac{1}{r^2} \dfrac{\partial}{\partial r} \left(r^2 \dfrac{\partial v_r}{\partial r} \right) + \dfrac{1}{r^2 \sin\theta} \dfrac{\partial}{\partial \theta} \left(\sin\theta \dfrac{\partial v_r}{\partial \theta} \right) + \dfrac{1}{r^2 \sin^2\theta} \dfrac{\partial^2 v_r}{\partial \phi^2} \right.$$
$$\left. - \dfrac{2}{r^2} v_r - \dfrac{2}{r^2} \dfrac{\partial v_\theta}{\partial \theta} - \dfrac{2}{r^2} v_\theta \cot\theta - \dfrac{2}{r^2 \sin\theta} \dfrac{\partial v_\phi}{\partial \phi} \right]$$

θ component: $\rho \left(\dfrac{\partial v_\theta}{\partial t} + v_r \dfrac{\partial v_\theta}{\partial r} + \dfrac{v_\theta}{r} \dfrac{\partial v_\theta}{\partial \theta} + \dfrac{v_\phi}{r \sin\theta} \dfrac{\partial v_\theta}{\partial \phi} + \dfrac{v_r v_\theta}{r} - \dfrac{v_\phi^2 \cot\theta}{r} \right) = -\dfrac{1}{r} \dfrac{\partial \mathcal{P}}{\partial \theta}$

$$+ \eta \left[\dfrac{1}{r^2} \dfrac{\partial}{\partial r} \left(r^2 \dfrac{\partial v_\theta}{\partial r} \right) + \dfrac{1}{r^2 \sin\theta} \dfrac{\partial}{\partial \theta} \left(\sin\theta \dfrac{\partial v_\theta}{\partial \theta} \right) + \dfrac{1}{r^2 \sin^2\theta} \dfrac{\partial^2 v_\theta}{\partial \phi^2} + \dfrac{2}{r^2} \dfrac{\partial v_r}{\partial \theta} \right.$$
$$\left. - \dfrac{v_\theta}{r^2 \sin^2\theta} - \dfrac{2\cos\theta}{r^2 \sin^2\theta} \dfrac{\partial v_\phi}{\partial \phi} \right]$$

ϕ component: $\rho \left(\dfrac{\partial v_\phi}{\partial t} + v_r \dfrac{\partial v_\phi}{\partial r} + \dfrac{v_\theta}{r} \dfrac{\partial v_\phi}{\partial \theta} + \dfrac{v_\phi}{r \sin\theta} \dfrac{\partial v_\phi}{\partial \phi} + \dfrac{v_r v_\phi}{r} + \dfrac{v_\phi v_\theta}{r} \cot\theta \right) = -\dfrac{1}{r \sin\theta} \dfrac{\partial \mathcal{P}}{\partial \phi}$

$$+ \eta \left[\dfrac{1}{r^2} \dfrac{\partial}{\partial r} \left(r^2 \dfrac{\partial v_\phi}{\partial r} \right) + \dfrac{1}{r^2 \sin\theta} \dfrac{\partial}{\partial \theta} \left(\sin\theta \dfrac{\partial v_\phi}{\partial \theta} \right) + \dfrac{1}{r^2 \sin^2\theta} \dfrac{\partial^2 v_\phi}{\partial \phi^2} - \dfrac{v_\phi}{r^2 \sin^2\theta} \right.$$
$$\left. + \dfrac{2}{r^2 \sin\theta} \dfrac{\partial v_r}{\partial \phi} + \dfrac{2\cos\theta}{r^2 \sin^2\theta} \dfrac{\partial v_\theta}{\partial \phi} \right]$$

Note that these equations are obtained formally from the Cauchy momentum and Navier-Stokes equations in Tables 2.3 and 2.4, respectively, by setting $\rho = 0$. This is consistent with the notion that mass (density) enters the momentum balance only through the inertial terms.

It is interesting to note that terms involving rates of change with time no longer appear in the momentum equation in the creeping flow approximation. Hence, this is a *quasi-steady-state* approximation, where time dependence does not appear explicitly. This is an important simplification, as we shall see when we consider the modeling of compression molding.

To see how we might use the momentum equation, let us return again to the analysis of flow between converging planes, shown in Figure 2.1. We have already seen (Equation 2.5) that $r v_r = \phi(\theta)$, which is a consequence of our assumption that $v_z = v_\theta = 0$ (purely radial flow). We will assume we have a Newtonian liquid in creeping flow, so we use the equations for cylindrical coordinates in Table 2.4 with $\rho = 0$. When we substitute the given form of the velocity into the equations, we find that most terms are identically zero, and the equations simplify to the following:

$$r: 0 = -\frac{\partial \mathcal{P}}{\partial r} + \frac{\eta}{r^3} \frac{d^2 \phi}{d\theta^2}, \tag{2.17a}$$

$$\theta: 0 = -\frac{1}{r} \frac{\partial \mathcal{P}}{\partial \theta} + \frac{2\eta}{r^3} \frac{d\phi}{d\theta}, \tag{2.17b}$$

$$z: 0 = -\frac{\partial \mathcal{P}}{\partial z}. \tag{2.17c}$$

Equation 2.17c simply tells us that the equivalent pressure does not depend on z, which is consistent with the assumption that z is a neutral direction.

Now, Equations 2.17a and b contain two variables, $\mathcal{P}$ and $\phi(\theta)$; $\mathcal{P}$ may depend on both r and θ. There are several ways to solve this pair of equations in two variables, but it is not our goal here to discuss methods of solving partial differential equations. The simplest (but least elegant) approach is to eliminate the pressure between the two equations to obtain a single equation for $\phi(\theta)$. To do this we use the fact that $\partial^2 \mathcal{P}/\partial r \partial \theta = \partial^2 \mathcal{P}/\partial \theta \partial r$; that is, the order of differentiation does not matter for "smooth" functions. We multiply Equation 2.17b by r so that we can write the two equations as follows:

$$\frac{\partial \mathcal{P}}{\partial r} = \frac{\eta}{r^3} \frac{d^2 \phi}{d\theta^2}, \tag{2.18a}$$

$$\frac{\partial \mathcal{P}}{\partial \theta} = \frac{2\eta}{r^2} \frac{d\phi}{d\theta}. \tag{2.18b}$$

Equation 2.18a is valid at every r and θ, so we can differentiate both sides with respect to θ to obtain

$$\frac{\partial^2 \mathcal{P}}{\partial \theta \partial r} = \frac{\eta}{r^3} \frac{d^3 \phi}{d\theta^3}. \tag{2.19a}$$

Similarly, we differentiate Equation 2.18b with respect to r to obtain

$$\frac{\partial^2 \mathcal{P}}{\partial r \partial \theta} = -\frac{4\eta}{r^3}\frac{d\phi}{d\theta}. \tag{2.19b}$$

The right sides of these equations must be equal, so we obtain, finally,

$$\frac{d^3\phi}{d\theta^3} + 4\frac{d\phi}{d\theta} = 0. \tag{2.20}$$

Equation 2.20 is a linear ordinary differential equation for $\phi(\theta)$ with constant coefficients whose solution is

$$\phi(\theta) = C_1 + C_2 \sin 2\theta + C_3 \cos 2\theta. \tag{2.21}$$

C_1, C_2, and C_3 are constants of integration. The mathematical reason they arise is that we have performed three indefinite integrations in going from $d^3\phi/d\theta^3$ to $\phi(\theta)$, but the physical reason is what interests us: To complete the description of the flow problem, we must be able to specify three things about the flow at the *boundaries* of the region. We shall return to this important topic subsequently.

2.2.5 Conservation of Energy

The equation of conservation of energy for an incompressible, nonreactive, single-phase liquid in rectangular Cartesian coordinates is

$$\rho c_p \frac{DT}{Dt} \equiv \rho c_p \left(\frac{\partial T}{\partial t} + v_x \frac{\partial T}{\partial x} + v_y \frac{\partial T}{\partial y} + v_z \frac{\partial T}{\partial z} \right) = \kappa \left(\frac{\partial^2 T}{\partial x^2} + \frac{\partial^2 T}{\partial y^2} + \frac{\partial^2 T}{\partial z^2} \right) + \Phi$$

$$= \kappa \nabla^2 T + \Phi. \tag{2.22}$$

The heat capacity at constant pressure, c_p, may be temperature dependent, but we have assumed that the thermal conductivity, κ, is a constant. Φ is the dissipation function, which has the following form in rectangular Cartesian coordinates:

$$\Phi = \tau_{xx}\frac{\partial v_x}{\partial x} + \tau_{yy}\frac{\partial v_y}{\partial y} + \tau_{zz}\frac{\partial v_z}{\partial z} + \tau_{xy}\left(\frac{\partial v_x}{\partial y} + \frac{\partial v_y}{\partial x} \right) + \tau_{xz}\left(\frac{\partial v_x}{\partial z} + \frac{\partial v_z}{\partial x} \right)$$

$$+ \tau_{yz}\left(\frac{\partial v_y}{\partial z} + \frac{\partial v_z}{\partial y} \right). \tag{2.23}$$

Note that Φ is a sum of squares for a Newtonian fluid, shown in Equation 2.11. The energy equation and the dissipation function, with the latter specialized for incompressible Newtonian fluids, are given in Tables 2.5 and 2.6, respectively, for the major coordinate systems.

Equation 2.22 is a bit difficult to interpret physically because it is an equation for the change in temperature, not the change in energy, from which it is derived. Roughly, the left side represents the accumulation of the internal energy in a fluid element. The term with the thermal conductivity κ represents the heat flow to and from the fluid element because of thermal conduction, while the viscous dissipation term reflects the rate of increase in internal energy because of work done on

Table 2.5. *Energy equation for an incompressible fluid with constant thermal conductivity*

$$\rho c_p \frac{DT}{Dt} = \kappa \nabla^2 T + \Phi$$

Rectangular Cartesian (x, y, z) coordinates:

$$\rho c_p \left(\frac{\partial T}{\partial t} + v_x \frac{\partial T}{\partial x} + v_y \frac{\partial T}{\partial y} + v_z \frac{\partial T}{\partial z} \right) = \kappa \left(\frac{\partial^2 T}{\partial x^2} + \frac{\partial^2 T}{\partial y^2} + \frac{\partial^2 T}{\partial z^2} \right) + \Phi$$

Cylindrical (r, θ, z) coordinates:

$$\rho c_p \left(\frac{\partial T}{\partial t} + v_r \frac{\partial T}{\partial r} + \frac{v_\theta}{r} \frac{\partial T}{\partial \theta} + v_z \frac{\partial T}{\partial z} \right) = \kappa \left[\frac{1}{r} \frac{\partial}{\partial r} \left(r \frac{\partial T}{\partial r} \right) + \frac{1}{r^2} \frac{\partial^2 T}{\partial \theta^2} + \frac{\partial^2 T}{\partial z^2} \right] + \Phi$$

Spherical (r, θ, ϕ) coordinates:

$$\rho c_p \left(\frac{\partial T}{\partial t} + v_r \frac{\partial T}{\partial r} + \frac{v_\theta}{r} \frac{\partial T}{\partial \theta} + \frac{v_\phi}{r \sin \theta} \frac{\partial T}{\partial \phi} \right) = \kappa \left[\frac{1}{r^2} \frac{\partial}{\partial r} \left(r^2 \frac{\partial T}{\partial r} \right) + \frac{1}{r^2 \sin \theta} \frac{\partial}{\partial \theta} \left(\sin \theta \frac{\partial T}{\partial \theta} \right) + \frac{1}{r^2 \sin \theta} \frac{\partial^2 T}{\partial \phi^2} \right] + \Phi$$

the fluid element by the surrounding fluid. The "p–V" work associated with pushing fluid elements into and out of the system is partially hidden in the fact that the heat capacity at constant pressure, c_p, appears in the equation in place of the heat capacity at constant volume, c_v. The two heat capacities are essentially equal for liquids because of the near incompressibility, but there are delicate issues that do not concern us here in deriving energy equations. The literature is full of analyses using incorrectly derived energy equations, and it is always a challenge when using commercial software to determine that the energy equation being employed is correct. (Elegant graphics to show the solution to an incorrect equation is of little use.)

2.2.6 Boundary Conditions

The conservation equations are *differential equations*, and their solution requires integration, hence, unknown constants of integration. This is illustrated in

Table 2.6. *Dissipation function (Φ/η) for an incompressible Newtonian fluid and one half the second invariant of the deformation rate ($\frac{1}{2}II_D$) for any fluid*

Rectangular Cartesian (x, y, z) coordinates:

$$2 \left[\left(\frac{\partial v_x}{\partial x} \right)^2 + \left(\frac{\partial v_y}{\partial y} \right)^2 + \left(\frac{\partial v_z}{\partial z} \right)^2 \right] + \left[\frac{\partial v_y}{\partial x} + \frac{\partial v_x}{\partial y} \right]^2 + \left[\frac{\partial v_z}{\partial y} + \frac{\partial v_y}{\partial z} \right]^2 + \left[\frac{\partial v_x}{\partial z} + \frac{\partial v_z}{\partial x} \right]^2$$

Cylindrical (r, θ, z) coordinates:

$$2 \left[\left(\frac{\partial v_r}{\partial r} \right)^2 + \left(\frac{1}{r} \frac{\partial v_\theta}{\partial \theta} + \frac{v_r}{r} \right)^2 + \left(\frac{\partial v_z}{\partial z} \right)^2 \right] + \left[r \frac{\partial}{\partial r} \left(\frac{v_\theta}{r} \right) + \frac{1}{r} \frac{\partial v_r}{\partial \theta} \right]^2 + \left[\frac{1}{r} \frac{\partial v_z}{\partial \theta} + \frac{\partial v_\theta}{\partial z} \right]^2 + \left[\frac{\partial v_r}{\partial z} + \frac{\partial v_z}{\partial r} \right]^2$$

Spherical (r, θ, ϕ) coordinates:

$$2 \left[\left(\frac{\partial v_r}{\partial r} \right)^2 + \left(\frac{1}{r} \frac{\partial v_\theta}{\partial \theta} + \frac{v_r}{r} \right)^2 + \left(\frac{1}{r \sin \theta} \frac{\partial v_\phi}{\partial \phi} + \frac{v_r}{r} + \frac{v_\theta \cot \theta}{r} \right)^2 \right] + \left[r \frac{\partial}{\partial r} \left(\frac{v_\theta}{r} \right) + \frac{1}{r} \frac{\partial v_r}{\partial \theta} \right]^2$$

$$+ \left[\frac{\sin \theta}{r} \frac{\partial}{\partial r} \left(\frac{v_\phi}{\sin \theta} \right) + \frac{1}{r \sin \theta} \frac{\partial v_\theta}{\partial \phi} \right]^2 + \left[\frac{1}{r \sin \theta} \frac{\partial v_r}{\partial \phi} + r \frac{\partial}{\partial r} \left(\frac{v_\phi}{r} \right) \right]^2$$

Equation 2.21 for the θ-dependent portion of the velocity field in the converging flow problem, where we must evaluate the constants C_1, C_2, and C_3. Differential equations apply *within* a spatial domain, while constants of integration are determined by conditions at boundaries. Typical boundary conditions for fluid flow problems are the specifications of velocities, stresses, and overall flow rates, while for the energy equation we typically specify temperatures or heat fluxes at boundaries.

The most common boundary condition in fluid mechanics is the *no-slip* condition, which states that the fluid in contact with a solid surface has a velocity equal to that of the surface. If the surface is stationary, as in our converging flow example, then the fluid must have a zero velocity at the surface. This condition is not intuitively obvious; Navier (1823), for example, one of the founders of the modern discipline of fluid mechanics, believed that there should be a relative velocity between the fluid and the adjacent solid surface that is proportional to the wall shear stress. No-slip is now generally accepted as the appropriate condition for problems in fluid mechanics, but there is considerable evidence that the no-slip condition fails for at least some polymer melts and concentrated solutions at high stress levels. We shall return to this issue in Chapter 12.

For the converging flow problem, the no-slip condition requires $v_r = \phi(\theta)/r$ to vanish at $\theta = \pm\alpha$ or, equivalently, $\phi(\alpha) = \phi(-\alpha) = 0$. From Equation 2.21,

$$\phi(\alpha) = C_1 + C_2 \sin 2\alpha + C_3 \cos 2\alpha = 0, \tag{2.24a}$$

$$\phi(-\alpha) = C_1 - C_2 \sin 2\alpha + C_3 \cos 2\alpha = 0. \tag{2.24b}$$

It readily follows that $C_2 = 0$, $C_1 = -C_3 \cos 2\alpha$, and

$$\phi(\theta) = C_3 \left(\cos 2\theta - \cos 2\alpha \right) \tag{2.25}$$

The constant C_3 must still be determined, and we note that the flow rate has not yet been specified. Let $q = Q/L$ denote the volumetric flow rate per unit length into the paper. To calculate the flow rate per unit length through the surface A in Figure 2.2, we first note that the differential flow rate through the differential arc denoted "$d\theta$" is $v_r r d\theta$ (velocity $\times$ arc length); the total flow rate per unit area is then

$$q = \int_{-\alpha}^{\alpha} v_r r d\theta = \int_{-\alpha}^{\alpha} \frac{\phi(\theta)}{r} r d\theta = \int_{-\alpha}^{\alpha} \phi(\theta) d\theta = C_3 \int_{-\alpha}^{\alpha} \left(\cos 2\theta - \cos 2\alpha \right) d\theta$$

$$= C_3 \left(\sin 2\alpha - 2\alpha \cos 2\alpha \right). \tag{2.26}$$

q is the same at all radial positions because of incompressibility, so Equation 2.26 can be solved to give C_3, from which we obtain the velocity at all r and θ as

$$v_r = \frac{q}{r} \frac{\cos 2\theta - \cos 2\alpha}{\sin 2\alpha - 2\alpha \cos 2\alpha}. \tag{2.27}$$

Note that the velocity is negative (flow is in the direction of decreasing r), with the maximum magnitude along the center plane and a zero velocity at the walls. The

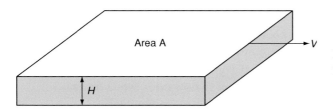

Area A

H

Figure 2.4. Schematic of a shearing experiment.

shear stress $\tau_{\theta r}$ is computed with the help of Table 2.3:

$$\tau_{\theta r} = \eta \left[r \frac{\partial}{\partial r} \left(\frac{v_\theta}{r} \right) + \frac{1}{r} \frac{\partial v_r}{\partial \theta} \right] = \frac{\eta}{r} \frac{\partial v_r}{\partial \theta} = -\frac{2\eta q}{r^2} \frac{\sin 2\theta}{\sin 2\alpha - 2\alpha \cos 2\alpha}. \qquad (2.28)$$

The shear stress $\tau_{\theta r}$ vanishes on the plane of symmetry, $\theta = 0$. The magnitude of the shear stress at the wall, $\theta = \pm \alpha$, is

$$\left| \tau_{\theta r}(r, \theta = \pm \alpha) \right| = \frac{2\eta q}{r^2} \frac{\sin 2\alpha}{\sin 2\alpha - 2\alpha \cos 2\alpha} = \frac{2\eta q}{r^2} \frac{1}{1 - 2\alpha \cot 2\alpha}. \qquad (2.29)$$

It is a useful exercise to consider Navier's boundary condition, in which there is a relative velocity between the fluid and the wall that is proportional to the shear stress. We write this relation as

$$\eta v_r = b \tau_{\theta r} \quad \text{at } \theta = \pm \alpha. \qquad (2.30)$$

b has dimensions of length and would presumably be a parameter that has a characteristic value for each liquid–solid pair. Using $v_r = \phi(\theta)/r$ and $\tau_{r\theta} = (\eta/r)\partial v_r/\partial \theta$ then gives $\eta\phi/r = \eta b\phi'/r^2$, where $\phi' = d\phi/d\theta$, or

$$\phi = \frac{b}{r} \frac{d\phi}{d\theta} \quad \text{at } \theta = \pm \alpha. \qquad (2.31)$$

There is clearly an inconsistency because this boundary condition would require that ϕ be a function of r, which is contrary to the assumption of radial flow. Hence, we can conclude that *purely radial flow is impossible if the fluid exhibits wall "slip."* (The exception is the case of complete slip, or $b \to \infty$, which corresponds to $\tau_{r\theta} = 0$ at $\theta = \pm \alpha$. It readily follows for this case that $v_r = q/2\alpha r$; that is, the velocity is the same at all angles and the fluid experiences no shear stress.)

2.3 Viscosity

Every fluid is characterized by a viscosity, which is a quantitative measure of the resistance to flow. Viscosity is best defined in terms of an experiment. Consider two very large ("infinite") parallel plates, as shown in Figure 2.4. The upper plate is moved relative to the lower plate with a velocity V. If we ignore the very small region near the edges of the plates, which are assumed to be at infinity, it is reasonable to assume that fluid elements move only in the same direction as the moving surface; in terms of the components v_x, v_y, and v_z of the velocity vector $\mathbf{v}$, this means that $v_y = v_z = 0$ and v_x depends only on y:

$$v_x = v_x(y), \quad v_y = v_z = 0. \qquad (2.32)$$

This velocity field clearly satisfies the continuity equation ($\partial v_x/\partial x + \partial v_y/\partial y + \partial v_z/\partial z = 0$). If we examine the terms in the Cauchy momentum equation, Table 2.2, we find that all terms on the left-hand side vanish at steady state ($\partial/\partial t = 0$). In this infinite geometry there is no preferred origin, so any x position in the plane should be equivalent to any other; this implies that stresses cannot vary with x, in which case $\partial \tau_{xx}/\partial x = 0$. Similarly, z is a neutral direction in which nothing varies, so $\partial \tau_{zx}/\partial z = 0$. (In fact, we expect $\tau_{zx} = 0$, since there is no shearing taking place along a z face.) Finally, we will take the direction of gravitational acceleration to be orthogonal to the planes, so $g_x = 0$. The x component of the momentum equation therefore simplifies to

$$\frac{\partial p}{\partial x} + \frac{\partial \tau_{yx}}{\partial y} = 0. \tag{2.33}$$

Now, we might do things to the flow field to cause a pressure gradient ($\partial p/\partial x$) in the flow direction, and we will consider one such important case subsequently in our analysis of the single-screw extruder. For this experiment, however, we assume that the plates are open to the atmosphere at both ends and there are no obstructions to the flow, in which case there is no reason to expect the pressure to vary from position to position. We therefore assume $\partial p/\partial x = 0$, in which case Equation 2.33 simplifies to $\partial \tau_{yx}/\partial y = 0$, or

$$\tau_{yx} = \text{constant.} \tag{2.34}$$

$\partial \tau_{yx}/\partial y = 0$ states only that τ_{yx} is independent of y, but we know from the infinite geometry that it must also be independent of x and z.

We have already defined a Newtonian fluid, and for this flow we simply obtain (Table 2.3)

$$\tau_{yx} = \eta \frac{dv_x}{dy} = \text{constant.} \tag{2.35}$$

If dv_x/dy is a constant, v_x must be linear; with the no-slip condition we have $v_x = 0$ at $y = 0$ and $v_x = V$ at $y = H$, so

$$v_x = V \frac{y}{H}, \tag{2.36a}$$

$$\frac{dv_x}{dy} = \frac{V}{H}. \tag{2.36b}$$

dv_x/dy is known as the *shear rate*, with dimensions of time^{-1}. We see from this result that if we measure the shear stress (force/area) required to move the plate as a function of shear rate (V/H), we will obtain a straight line for all V and H, and the slope will be the viscosity, η. Typical data are shown in Figure 2.5 for an oil used as a viscosity standard; the stress is measured in pascals (Pa), and the viscosity at 24 °C is approximately 0.03 pascal-seconds (Pa s).[*]

---

[*] While the pascal second is the accepted SI unit for viscosity, data are often reported in the CGS unit poise (P). 1 P = 0.1 Pa s, so the viscosity of the oil is 0.3 P, or 30 cP (centipoise). The viscosity of water at room temperature is about 1 cP = 1 mPa s (milli pascal second).

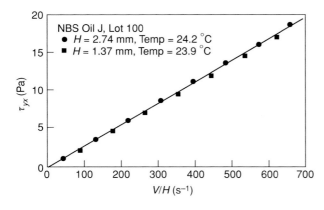

Figure 2.5. Shear stress as a function of shear rate, NBS Oil J, Lot 100, at approximately 24 °C.

Polymer melts rarely behave as Newtonian fluids, but we can still determine the viscosity with this experiment. Let us suppose that the stress depends on the deformation gradient in a completely arbitrary way. With the assumption we have made about the flow, the only nonvanishing part of the deformation gradient, regardless of how we define it, is dv_x/dy. Thus, while τ_{yx} may be an arbitrary function of dv_x/dy, the condition $\tau_{yx} =$ constant means that dv_x/dy must also be constant.[*] Thus, Equations 2.36a–b still follow, and the definition of the shear rate is unchanged.

The shear rate is commonly written $\dot{\gamma}$, although other notation is in use. Viscosity is defined as the ratio $\tau_{yx}/\dot{\gamma}$:

$$\eta\left(|\dot{\gamma}|\right) \equiv \frac{\tau_{yx}(\dot{\gamma})}{\dot{\gamma}}. \tag{2.37}$$

The viscosity is in general a function of the shear rate; the argument used here shows that viscosity is a *unique* function of shear rate for a given fluid at fixed temperature and pressure. Viscosity functions for some typical polymer melts are shown in Figure 1.7. Data are typically plotted on log–log coordinates, and, as noted in Chapter 1, two features are commonly observed: There is an approach to a region of constant viscosity at low shear rates (denoted η_o, the zero-shear viscosity), and there is a region of power-law behavior (a straight line on a logarithmic plot) at intermediate shear rates. There may also be a region of constant viscosity at high shear rates, but this region is rarely observed experimentally. Because the viscosity functions of polymer melts decrease with increasing shear rate, melts are often referred to as *shear thinning*.

There are molecular theories relating the viscosity of a linear polymer melt to chain length, but the most useful viscosity equations are empirical. Data are often represented by a power law (sometimes called the *Ostwald-de Waele* model):

$$\tau_{yx} = K\left|\dot{\gamma}\right|^{n-1}\dot{\gamma}, \tag{2.38a}$$

$$\eta = K\left|\dot{\gamma}\right|^{n-1}. \tag{2.38b}$$

---

[*] We are assuming that τ_{yx} is a monotonic function of dv_x/dy, so there is a unique solution to the equation $\tau_{yx}(dv_x/dy) =$ constant. The argument is essentially unchanged if multiple solutions are possible, but some interesting physical issues arise in that case.

This empirical equation may be a good fit to data over several decades in shear rate. The absolute value of the shear rate is needed to ensure that shear stress and shear rate, both of which are directional quantities, remain collinear, and that the viscosity is always real and positive. (It can be shown from the second law of thermodynamics that η must be non negative.) The *consistency index K* is strongly temperature dependent, while n is usually insensitive to temperature (although the shear rate range over which power-law behavior is observed will change with temperature).

The empirical *Carreau-Yasuda* (or, sometimes, *Carreau-Yasuda-Elbirli*) *model* contains two additional parameters and is extensively used to correlate melt data:

$$\eta = \eta_o \left(1 + \beta \, |\dot{\gamma}|^a \right)^{\frac{n-1}{a}} . \tag{2.39}$$

This equation goes to a zero-shear viscosity as $\dot{\gamma} \to 0$ and to a power law for $|\dot{\gamma}| \gg \beta^{-a}$; in the latter case, $\eta_0 \beta^{\frac{n-1}{a}}$ corresponds to the consistency index, K. The parameters appear to correlate with changes in molecular structure for polyolefins. The *Cross equation*, in which $a = 1 - n$, works well for many polymers, including polyesters and polyacetals.

We need to recognize that we have defined the viscosity entirely in terms of the response of the fluid to a single experiment. Other shearing geometries (flow between a rotating cone and plate, flow between rotating parallel plates, pressure-driven flow in a long cylindrical or planar channel) can be shown to be mathematically equivalent to the experiment described here, so the scope of measuring techniques is much greater than suggested by the elementary analysis. It can be shown that the viscosity is a *material function* that is a general property of the material and independent of any particular experimental or processing geometry.

Finally, the shear rate as defined by Equation 2.36b is clearly the appropriate argument for the viscosity function only for one-dimensional flows like the one used here. We need a quantity that reduces to $|dv_x/dy|$ for the one-dimensional flow but is properly invariant to the way in which we choose to define our coordinate system. The appropriate function, which follows directly from the principles of matrix algebra, is one half the *second invariant* of the rate of deformation, which is usually denoted $\frac{1}{2}II_D$. $\frac{1}{2}II_D$ is shown in Table 2.6, where it is identical to the dissipation function Φ divided by η for the special case of Newtonian fluids. (It is important to keep in mind that the function Φ/η in Table 2.6 is the proper form for the dissipation *only* for a Newtonian fluid, whereas $\frac{1}{2}II_D$ is a universally valid definition that depends only on the velocity field.) For an arbitrary flow field, then, the power-law and Carreau-Yasuda equations would be written, respectively,

$$\eta = K \left(\tfrac{1}{2}II_D \right)^{\frac{n-1}{2}} , \tag{2.40a}$$

$$\eta = \eta_o \left(1 + \beta (\tfrac{1}{2}II_D)^{\frac{a}{2}} \right)^{\frac{n-1}{a}} . \tag{2.40b}$$

At this point we need to classify fluids into those that are *inelastic*, by which we mean that the stress at a given time depends only on the deformation rate at that time, and those that are *viscoelastic*, by which we mean that the stress at a given time depends on the past as well as the present deformation rate. Polymer melts are usually viscoelastic; Figure 1.9, for example, shows the evolution of the shear stress in a polyethylene melt at a constant shear rate over a time scale of several seconds. The viscosity at this shear rate is determined from the ultimate steady-state shear stress, but the shear stress at any time during the transient clearly depends on the history of deformation. Despite this dependence on prior deformation, we can gain considerable insight into melt processing by first considering inelastic liquids (and an inelastic description of the stress is frequently adequate, especially for some polyesters, nylons, and polycarbonates). The extra stress for incompressible inelastic liquids is as given in Table 2.3, but the viscosity η is a function of $\frac{1}{2}II_D$ – as given, for example, by Equation 2.40a or 2.40b. It is important to recall that substitution of this extra stress with a variable viscosity into the Cauchy momentum equation does not lead to the Navier-Stokes equations.

BIBLIOGRAPHICAL NOTES

The continuity and momentum equations are derived in most of the textbooks on fluid mechanics; see, for example,

Denn, M. M., *Process Fluid Mechanics*, Prentice Hall, Englewood Cliffs, NJ, 1980.

For an elegant and rigorous treatment, see

Aris, R., *Vectors, Tensors, and the Basic Equations of Fluid Dynamics*, Prentice Hall, Englewood Cliffs, NJ, 1962; reprinted by Dover Publications, Mineola, NY, 1990.

The development of the correct energy equation is delicate and is often done incorrectly. Correct derivations may be found in Aris and in

Bird, R. B., W. E. Stewart, and E. N. Lightfoot, *Transport Phenomena,* 2nd ed., John Wiley, New York, 2006.

There is a catalogue of applications of incorrect energy balances in

Denn, M. M., *Process Modeling*, Longman Scientific and John Wiley, New York, 1986, Ch. 5.

For an analysis of converging flow with a Navier slip condition for Newtonian, power-law, and Carreau-Yasuda fluids, see

Joshi, Y. M., and M. M. Denn, *J. Non-Newtonian Fluid Mech.*, **114**, 185 (2003).

3 Extrusion

3.1 Introduction

The extruder, shown schematically in Figure 1.1, is central to most melt processing operations. We can achieve considerable insight into the operation and design of single-screw extruders by remarkably simple models, despite the mechanical complexity. We begin this chapter by obtaining velocity, stress, and temperature distributions for flow in straight channels with parallel walls of "infinite" length. The infinite channel results are important in and of themselves, but we shall see here that they lead immediately to a model for the single-screw extruder as well. The results also provide an important framework for the modeling of flows in situations in which the walls are not parallel, which we address in Chapter 5.

3.2 Plane Channel

3.2.1 Stress Distribution

Let us suppose we have steady isothermal flow (i.e., the temperature is constant throughout the flow field and all $\partial/\partial t = 0$) between two infinite parallel planes, as shown in Figure 3.1. The flow is in the x direction. We assume for generality that there is a finite pressure gradient ($\partial p/\partial x \neq 0$) and that the surface at $y = 0$ moves relative to the surface at $y = H$ with a constant velocity V. We shall see subsequently that the results obtained here will form the foundation for the modeling of single-screw extrusion and the extrusion coating of flat sheets.

If we place a drop of colored dye in the flow field, we will find that the dye moves parallel to the wall. This is expected, since the walls are parallel and there is no driving force to cause fluid to move orthogonal to the walls or into the plane of the paper. Thus, we may assume that the Cartesian components of the velocity vector may be written

$$v_x = v_x(y), \, v_y = v_z = 0. \tag{3.1}$$

Figure 3.1. Flow between infinite parallel planes.

(We know that v_x must be nonzero because we are assuming that there is a net flow in the x direction. v_x must be a function of y because the no-slip condition requires $v_x = V$ at $y = 0$ and $v_x = 0$ at $y = H$.) This velocity field automatically satisfies the continuity equation, $\partial v_x/\partial x + \partial v_y/\partial y + \partial v_z/\partial z = 0$. We do need to keep in mind that the velocity field might be more complex near the beginning and end of the flow field, where fluid enters and leaves the channel, but for now we assume that these positions are located an infinite distance away and do not affect the flow.

With the assumed velocity field, all terms on the left side of the three components of the Cauchy momentum equation, Equations 2.6a–c and Table 2.2, are zero. The equations in terms of the equivalent pressure $\mathcal{P}$ thus simplify to

$$x \text{ component:} \quad 0 = -\frac{\partial \mathcal{P}}{\partial x} + \frac{\partial \tau_{xx}}{\partial x} + \frac{\partial \tau_{yx}}{\partial y} + \frac{\partial \tau_{zx}}{\partial z}, \tag{3.2a}$$

$$y \text{ component:} \quad 0 = -\frac{\partial \mathcal{P}}{\partial y} + \frac{\partial \tau_{xy}}{\partial x} + \frac{\partial \tau_{yy}}{\partial y} + \frac{\partial \tau_{zy}}{\partial z}, \tag{3.2b}$$

$$z \text{ component:} \quad 0 = -\frac{\partial \mathcal{P}}{\partial z} + \frac{\partial \tau_{xz}}{\partial x} + \frac{\partial \tau_{yz}}{\partial y} + \frac{\partial \tau_{zz}}{\partial z}. \tag{3.2c}$$

This is still a formidable-looking set of equations, but they simplify considerably more. The z direction is a "neutral" direction in which nothing is happening (all z planes are identical), so all $\partial/\partial z = 0$. There is no preferred origin in this infinite geometry, so all x positions must be equivalent as far as the stress and flow field are concerned. Thus, even though we have not specified a particular constitutive relation for the stress, we can conclude that stress will not change with position in the x direction, in which case $\partial \tau_{xx}/\partial x = \partial \tau_{xy}/\partial x = \partial \tau_{xz}/\partial x = 0$. Finally, because there is no relative motion in the z direction between adjacent xy planes, there can be no shear stress $\tau_{xz} = \tau_{zx}$. Thus, our equations reduce to

$$0 = -\frac{\partial \mathcal{P}}{\partial x} + \frac{\partial \tau_{yx}}{\partial y}, \tag{3.3a}$$

$$0 = -\frac{\partial \mathcal{P}}{\partial y} + \frac{\partial \tau_{yy}}{\partial y} = \frac{\partial}{\partial y}\left(-\mathcal{P} + \tau_{yy}\right), \tag{3.3b}$$

$$0 = -\frac{\partial \mathcal{P}}{\partial z}. \tag{3.3c}$$

Equation 3.3c simply tells us that the equivalent pressure does not change in the z direction, which is obvious. Equation 3.3b tells us that the sum $\sigma_{yy} = -\mathcal{P} + \tau_{yy}$ is independent of y, so we can write

$$\mathcal{P} = \tau_{yy} + \text{function of } x. \tag{3.4}$$

Now, τ_{yy} might depend on y but, like the other stresses, it cannot depend on the flow direction, x. Thus, $\mathcal{P}$ might depend on both x and y, but only in the form of a sum of a function of x and a function of y; hence, $\partial\mathcal{P}/\partial x$ will depend only on x. We are now in a position to solve our flow problem.

We rewrite Equation 3.3a as

$$\frac{\partial \mathcal{P}}{\partial x} = \frac{d\tau_{yx}}{dy}. \tag{3.5}$$

We use the ordinary derivative d/dy on the right because τ_{yx} can be a function only of y. The left-hand side of Equation 3.5 is a function only of x, whereas the right-hand side is a function only of y. x and y are independent variables, and we are free to change one while keeping the other constant. If we change x but keep y constant, then $d\tau_{yx}/dy$ cannot change, hence, neither can $\partial\mathcal{P}/\partial y$; thus, $\partial\mathcal{P}/\partial x$ must be independent of x, since it cannot change when we change x. By a similar argument, $d\tau_{yx}/dy$ must be independent of y. We can therefore write

$$\frac{\partial \mathcal{P}}{\partial x} = \frac{d\tau_{yx}}{dy} = \text{constant}. \tag{3.6}$$

We therefore conclude that for *any* fluid, the equivalent pressure varies linearly in the flow direction and the shear stress varies linearly across the flow channel. Since $\partial\mathcal{P}/\partial x$ is a constant, we can integrate Equation 3.6 with respect to y to obtain

$$\tau_{yx} = \frac{\partial \mathcal{P}}{\partial x} y + C_1, \tag{3.7}$$

where C_1 is a constant of integration.

3.2.2 Newtonian Fluid

We can now obtain the velocity distribution for any viscosity function – the power law given by Equation 2.38, for example. For analytical simplicity we begin with the Newtonian fluid, in which case Equation 3.7 becomes

$$\tau_{yx} = \eta \frac{dv_x}{dy} = \frac{\partial \mathcal{P}}{\partial x} y + C_1. \tag{3.8}$$

Since the derivative of the velocity is linear, the velocity itself must be quadratic. Integration of both sides of Equation 3.8 thus yields

$$v_x(y) = \frac{1}{2\eta} \frac{\partial \mathcal{P}}{\partial x} y^2 + \frac{C_1}{\eta} y + C_2. \tag{3.9}$$

The constants C_1 and C_2 are determined from the conditions $v_x = V$ at $y = 0$ and $v_x = 0$ at $y = H$:

$$v_x(0) = V = C_2, \tag{3.10a}$$

$$v_x(H) = 0 = \frac{1}{2\eta} \frac{\partial \mathcal{P}}{\partial x} H^2 + \frac{C_1}{\eta} H + C_2. \tag{3.10b}$$

Figure 3.2. Schematic of an elementary plane extruder.

Solving for C_1 and C_2, we thus obtain

$$v_x(y) = \frac{H^2}{2\eta} \frac{\partial \mathcal{P}}{\partial x} \frac{y}{H} \left(\frac{y}{H} - 1 \right) + V \left(1 - \frac{y}{H} \right). \tag{3.11}$$

In the absence of a pressure gradient ($\partial \mathcal{P}/\partial x = 0$) we obtain a linear velocity profile, as in Section 2.3. If both plates are stationary ($V = 0$) the velocity profile is a parabola that is symmetric about the center plane $y = H/2$, and the shear stress vanishes at the center plane.

The flow rate in the x direction per unit width, q, is found by integration of the velocity:

$$q = \int_0^H v_x(y)dy = -\frac{H^3}{12\eta} \frac{\partial \mathcal{P}}{\partial x} + \frac{VH}{2}. \tag{3.12}$$

There are two contributions to the overall flow, the *drag flow* proportional to V and the *pressure-driven flow* proportional to the pressure gradient. Note that if the pressure at the channel entrance is greater than the pressure at the exit, the pressure decreases along the channel length, so $\partial \mathcal{P}/\partial x < 0$ and the pressure term contributes to a flow in the positive x direction.

3.3 An Elementary Extruder

3.3.1 Design Equation

Now consider the flow geometry shown in Figure 3.2. We again have two infinite plates, with the lower plate moving with relative velocity V, but we now suppose that the channel is obstructed at a position $x = L$, where it is connected to a channel of height $h < H$ and length l. (There are clearly mechanical problems in implementing this flow, which we shall ignore for the present.) There is a fluid reservoir at $x = 0$ that is maintained at atmospheric pressure, which we can take without loss of generality to be $p = 0$, and the exit of the small channel is also assumed to be at atmospheric pressure.

Let us first consider the small channel of width h and length l. We assume $l \gg h$, so the flow rearrangements near the beginning and the end of the channel are not important to the analysis of the overall behavior. We will ignore differences between p and $\mathcal{P}$ in this analysis because we assume h and H are both small and the gravitational contribution across the channel is negligible. There is no moving plane in the small channel, so the drag term in Equation 3.12 is zero and we may write

$$\frac{\mathcal{P}|_{x=L} - 0_{(atmospheric)}}{l} = \frac{12\eta q}{h^3}. \tag{3.13a}$$

We have made use of the fact that the pressure gradient is constant, which allows us to replace $\partial \mathcal{P}/\partial x$ with (final pressure – initial pressure)/length. Alternatively, we may write Equation 3.13a as

$$\mathcal{P}|_{x=L} = \frac{12\eta q l}{h^3}. \tag{3.13b}$$

This is the pressure required at $x = L$ to ensure that we will have a flow rate/unit width q in the x direction in the small channel. Thus, we must choose the wall velocity V in the large channel to ensure that we obtain this pressure. Note that we are asking for something counterintuitive, because we want to bring about a pressure *increase* in the flow direction in the large channel.

If we ignore the small regions around $x = 0$ and $x = L$, where the assumption that $v_x = v_x(y)$, $v_y = 0$ is clearly incorrect, Equations 3.11 and 3.12 should apply. Because the pressure gradient is constant, we can replace $\partial \mathcal{P}/\partial x$ by $(\mathcal{P}|_{x=L} - \mathcal{P}|_{x=0})/L$, or

$$\frac{\partial \mathcal{P}}{\partial x} = \frac{\mathcal{P}|_{x=L} - \mathcal{P}|_{x=0}}{L} = \frac{12\eta q l}{h^3 L}. \tag{3.14}$$

Equation 3.12 then becomes

$$q = -\left(\frac{H^3}{12\eta}\right)\left(\frac{12\eta q l}{h^3 L}\right) + \frac{V H}{2} \tag{3.15a}$$

or

$$V = \frac{2q}{H}\left[1 + \left(\frac{H}{h}\right)^3 \frac{l}{L}\right]. \tag{3.15b}$$

Equation 3.15b is the design equation for a simple extruder, which takes fluid from a reservoir and extrudes it at a rate q through a die of gap height h. The velocity profile away from the constriction and the reservoir is obtained by substituting Equations 3.14, 3.15a, and 3.15b into Equation 3.11 to obtain

$$\frac{v_x}{V} = \frac{3\beta}{(1 + \beta)}\xi(\xi - 1) + (1 - \xi), \tag{3.16a}$$

where

$$\xi = y/H, \quad \beta = (H/h)^3 l/L. \tag{3.16b,c}$$

From Equation 3.15b, β reflects the relative increase in wall velocity to achieve the required pressure *increase* from the reservoir in order to extrude the fluid through the small die. The velocity profile away from the constriction and the reservoir is shown in Figure 3.3; the pressure buildup causes a backflow, with the velocity vanishing at an intermediate position $y/H = (1 + \beta)/3\beta$ and becoming negative over a portion of the channel if $\beta > 1/2$. The ratio of pressure-driven backflow (q_p) to drag flow (q_d) is given by $-\beta/(1 + \beta)$; the maximum backflow is equal in magnitude

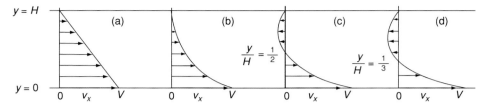

Figure 3.3. Velocity profiles in a plane channel. (a) $q_p/q_d = 0$, $\beta = 0$; (b) $q_p/q_d = -1/3$, $\beta = 1/2$; (c) $q_p/q_d = -2/3$, $\beta = 2$; (d) $q_p/q_d = -1$, $\beta = \infty$.

to the drag flow, which occurs in the limit $\beta \to \infty$. Clearly there will be a region near the constriction where the fluid carried forward by the drag flow reverses and returns with the pressure-driven backflow, but we cannot account for this flow with the simplified analysis used here.

Finally, we have not computed the cost of operating this device in terms of power. The power input per unit width in the z direction is simply the product of the wall shear stress, the length, and the wall velocity.* The shear stress at the wall is obtained from Equation 3.11 as

$$\tau_{yx}\big|_{y=0} = \frac{d}{dy}\left\{\left[\frac{H^2}{2}\frac{\partial \mathcal{P}}{\partial x}\frac{y}{H}\left(1-\frac{y}{H}\right)\right] + \eta V\left(1-\frac{y}{H}\right)\right\}\bigg|_{y=0} = -\frac{\eta}{H}\frac{2q}{H}(1+4\beta),$$
(3.17)

and the power per unit width in the z direction, $\wp/W$, is

$$\frac{\wp}{W} = \frac{4\eta L q^2}{H^3}(1+\beta)(1+4\beta),$$
(3.18)

where we have made use of all the intermediate equations. The power requirement for a fixed throughput scales quadratically when β is greater than unity.

EXAMPLE 3.1. Suppose we wish to produce 10 kg/hr of a polymer with $\rho = 782$ kg/m^3 and $\eta = 30,000$ Pa s through a slit die with a gap of $h = 2.5$ mm, $l/h = 30$, and a width $W = 1.5$ m. The dimensions of the upstream channel are $H = 20$ mm and $L = 0.75$ m. The volumetric throughput per unit width is

$$q = \left(10\frac{\text{kg}}{\text{hr}}\right)\left(\frac{1}{782\,\text{kg/m}^3}\right)\left(\frac{1}{1.5\,\text{m}}\right)\left(\frac{1}{3,600\,\text{s/hr}}\right) = 2.37 \times 10^{-6}\,\text{m}^2/\text{s}.$$

The average velocity from the die, q/h, is about 1 mm/s, which is very slow. We find the pressure at the entrance to the die (hence, the pressure buildup in the extruder and the pressure drop across the die) from Equation 3.13b as

$$\mathcal{P} = 12(3 \times 10^4\,\text{Pa s})(2.37 \times 10^{-6}\,\text{m}^2/\text{s})(75 \times 10^{-3}\,\text{m})/(2.5 \times 10^{-3}\,\text{m})^3$$
$$= 4.1 \times 10^6\,\text{Pa}$$

---

* Work = force × distance. Power = rate of doing work = force × rate of change of distance = force × velocity. Force = shear stress × area. Force/unit width = $\tau_{yx}L$. Power/unit width = $\tau_{yx}LV$.

The wall shear stress in the die is computed from the balance between pressure drop multiplied by cross-sectional area and wall shear stress multiplied by surface area as

$$\tau_w = \mathcal{P}h/2l = (4.1 \times 10^6 \, \text{Pa})(2.5 \times 10^{-3} \, \text{m})/2(75 \times 10^{-3} \, \text{m}) = 6.8 \times 10^4 \, \text{Pa}$$
$$= 0.068 \, \text{MPa}.$$

As we shall see subsequently, the maximum wall stress for most extrusion operations is in the neighborhood of 0.1 MPa, so this is a reasonable value.

β is given by Equation 3.16c:

$$\beta = \left(\frac{20}{2.5}\right)^3 \left(\frac{0.075}{0.75}\right) = 51.2.$$

The power requirement is then obtained from Equation 3.18 as

$$\wp = \frac{4(3 \times 10^4 \, \text{Pa s})(0.75 \, \text{m})}{(20 \times 10^{-3} \, \text{m})} \left(\frac{2.37 \times 10^{-6} \, \text{m}^2\text{/s}}{20 \times 10^{-3} \, \text{m}}\right)^2 (52.2)(1 + 4 \times 51.2)(1.5)$$
$$= 1{,}018 \, \text{W} \cong 1 \, \text{kW}.$$

Note that everything is computed using SI units, so no conversions are needed. 1 kW is about 1.3 hp. From Equation 3.15b we find the velocity of the moving wall to be

$$V = \frac{2 \times (2.37 \times 10^{-6} \, \text{m}^2\text{/s})}{20 \times 10^{-3} \, \text{m}}(52.2) = 1.24 \times 10^{-2} \, \text{m/s} = 12.4 \, \text{mm/s}.$$

3.3.2 Temperature Development

We have not addressed the issue of cooling, which may be necessary to keep the temperature from rising because of the power input and the viscous dissipation; this requires solution of Equation 2.22. The detailed design of heat transfer systems is beyond the scope of what we wish to accomplish in this text, but it is useful to sketch out a simplified analysis. We will assume that all physical properties are independent of temperature, thus ignoring the strong dependence of viscosity on temperature for polymer melts, which we address in Chapter 4. This assumption uncouples the momentum and energy equations and permits us to use the velocity and pressure fields that we have already calculated with the assumption of isothermality.

We make one important assumption here in applying Equation 2.22, namely, that energy transport in the flow direction by convection ($\rho c_p v_x \partial T/\partial x$) is far more important than transport in the flow direction by conduction ($\kappa \partial^2 T/\partial x^2$). The validity of this assumption follows from the small thermal conductivity of polymer melts; it can be established from a scaling argument for $\rho c_p q/\kappa \gg 1$, which is generally true.* The conduction term in the transverse direction, $\kappa \partial^2 T/\partial y^2$, must be retained; there is no transverse flow; hence, conduction is the only mechanism available to provide a heat flow to the walls, where the heat transfer takes place. The only nonzero term in the dissipation function is $\eta(dv_x/dy)^2$, so at steady state ($\partial T/\partial t = 0$) Equation 2.22 becomes

$$\rho c_p v_x \frac{\partial T}{\partial x} = \kappa \frac{\partial^2 T}{\partial y^2} + \eta \left(\frac{dv_x}{dy}\right)^2. \tag{3.19}$$

There will be one integration with respect to x and two with respect to y, so we will need to provide one piece of boundary information in the x direction and two in the y direction. The x condition appears to be straightforward: We assume that at $x = 0$ the melt is uniformly at the reservoir temperature, which we denote T_i (for *initial*). The thermal boundary condition at a wall is typically written as an equality between the heat flux into the wall from conduction in the fluid and the heat flux from the wall to the surrounding heat transfer medium. It is an equality because the wall is assumed to have no thermal capacitance, so the flux into the wall must equal the flux out. The heat flux in the fluid is equal to $-\kappa \partial T/\partial y$. (This is known as *Fourier's law*, but it is an empirical constitutive equation, not a law of nature.) The flux to the surroundings is usually written as $U(T - T_a)$, where T_a is the temperature of the ambient environment, which might be air or a heat exchange fluid. U is an *overall heat transfer coefficient*, which is characteristic of the particular geometry, materials, and flow. The appropriate boundary conditions are then

$$-\kappa \frac{\partial T}{\partial y} = U(T - T_a) \text{ at } y = 0, H, \tag{3.20}$$

where U may be different on the two surfaces. Equation 3.20 is often called *Newton's law of cooling*, but it is actually a definition of the heat transfer coefficient and not a law of nature.

Equation 3.19, with v_x given by Equation 3.16, does not lend itself to a simple analytical solution, but we can gain insight from an approximation. A quantity that often arises in heat transfer analysis is the *cup-mixing temperature*, defined as

$$T_{cm} = \frac{\int_0^H v_x T \, dy}{q}. \tag{3.21}$$

T_{cm} is the average temperature of an extrudate that has been collected and mixed, hence, the name "cup mixing." Note that T_{cm} depends only on x. We can express Equation 3.19 in terms of T_{cm} by integrating both sides with respect to y from zero to H, making use of the fact that the physical properties are all constant:

$$\rho c_p \int_0^H v_x \frac{\partial T}{\partial x} dy = \kappa \int_0^H \frac{\partial^2 T}{\partial y^2} dy + \eta \int_0^H \left(\frac{dv_x}{dy}\right)^2 dy. \tag{3.22}$$

---

* The grouping $\kappa/\rho c_p$ is known as the *thermal diffusivity*. The value for high- and low-density polyethylene at typical processing temperature is 1.2×10^{-7} m²/s, while the value for polystyrene is 0.8×10^{-7} m²/s. If we use the value $q = 2.4 \times 10^{-6}$ m²/s from Example 3.1, we then obtain $\rho c_p q/\kappa \sim 20\text{--}30$, which certainly satisfies the condition $\rho c_p q/\kappa \gg 1$.

Now, v_x is independent of x, so $v_x \partial T/\partial x = \partial(v_x T)/\partial x$, and, since H is a constant, we can interchange the order of integration with respect to y and differentiation with respect to x. Thus,

$$\rho c_p \int_0^H v_x \frac{\partial T}{\partial x} dy = \rho c_p \frac{d}{dx} \int_0^H v_x T \, dy = \rho c_p q \frac{dT_{cm}}{dx}. \tag{3.23a}$$

Also,

$$\kappa \int_0^H \frac{\partial^2 T}{\partial y^2} dy = \kappa \left[\frac{\partial T}{\partial y}\bigg|_{y=H} - \frac{\partial T}{\partial y}\bigg|_{y=0} \right]. \tag{3.23b}$$

Finally, from Equation 3.16a,

$$\frac{\partial v_x}{\partial y} = \frac{\partial v_x}{\partial \xi}\frac{\partial \xi}{\partial y} = \frac{1}{H}\frac{\partial v_x}{\partial \xi} = \frac{12\beta q}{H^3} y - \frac{2q}{H^2}(1+4\beta) = \alpha y - \gamma, \tag{3.23c}$$

where

$$\alpha = \frac{12\beta q}{H^3}, \quad \gamma = \frac{2q}{H^2}(1+4\beta). \tag{3.23d,e}$$

Thus, Equation 3.22 becomes

$$\rho c_p q \frac{dT_{cm}}{dx} = \kappa \left[\frac{\partial T}{\partial y}\bigg|_{y=H} - \frac{\partial T}{\partial y}\bigg|_{y=0} \right] + \eta \int_0^H (\alpha y - \gamma)^2 dy \tag{3.24}$$

or, making use of the boundary condition Equation 3.20 and carrying out the integration of the dissipative term,

$$\rho c_p q \frac{dT_{cm}}{dx} = -U_H(T_H - T_{aH}) + U_o(T_o - T_{ao}) + \frac{4\eta q^2}{H^3}(4\beta^2 + 2\beta + 1). \tag{3.25}$$

Subscripts H and O refer, respectively, to quantities evaluated at $y = H$ and $y = 0$.*

If we assume that both walls of this simple extruder are insulated, there will be no heat flux and $U_H = U_O = 0$. In that case, which we denote *adiabatic operation*, Equation 3.25 simplifies to

$$\frac{dT_{cm}}{dx} = \frac{4\eta q}{\rho c_p H^3}(4\beta^2 + 2\beta + 1) = \text{constant}. \tag{3.26}$$

Thus, the temperature increases linearly and can never reach a steady state, and the cup-mixing temperature at the outlet of the extruder (neglecting any further dissipation in the die) is

$$\text{adiabatic:} \quad T_{cm}(L) = T_i + \frac{4\eta q L}{\rho c_p H^3}(4\beta^2 + 2\beta + 1). \tag{3.27}$$

---

* Readers who have studied thermodynamics will note, by comparison of Equations 3.13b, 3.18, and 3.25, that

$$\wp/W = q\mathcal{P}|_{x=L} + L\int_0^H \Phi \, dy;$$

that is, the power input equals the sum of the total dissipation and the rate of doing "flow work."

Equation 3.27 will give a good estimate of the adiabatic temperature rise, hence, an upper bound on the effluent cup-mixing temperature, but it is important to note that it gives no details about the temperature *distribution* in the extruder in the transverse direction. Thus, we have no way of knowing the maximum temperature to which the polymer is exposed during the processing.

EXAMPLE 3.2. Now suppose we wish to calculate the adiabatic temperature rise, $T_{cm}(L) - T_i$, for the flow problem in Example 3.1. The specified melt density of 782 kg/m^3 is that of polyethylene. We will suppose we have a low-density polyethylene, for which the heat capacity at 150 °C is 2.57×10^3 J/(kg K); the value for high-density polyethylene is 2.65×10^3 J/(kg K), so the result will be essentially the same. We are given $\eta = 3 \times 10^4$ Pa s, $q = 2.37 \times 10^{-6}$ m^2/s, $L = 0.75$ m, $H = 0.2 \times 10^{-3}$ m, and $\beta = 51.2$. Substituting into Equation 3.27 then gives

$$T_{cm}(L) - T_i = \frac{4\eta q L}{\rho c_p H^3}(4\beta^2 + 2\beta + 1) = 139\ °C.$$

This is a large number, and the assumption of constant physical properties is sure to fail. Furthermore, if we assume that the feed temperature is 140 °C, which is 20–25 °C above the melting temperature of LDPE (cf. Table 1.1 and Figure 1.12), the average temperature will be 280 °C, at which polyethylene is likely to degrade.

It is more realistic to assume that there is a heat transfer medium on the stationary wall, but that the moving wall is adiabatic. (In a real situation we might have to consider conduction along the metallic moving wall in accounting for all heat transfer mechanisms.) Equation 3.25 then becomes

$$\rho c_p q \frac{dT_{cm}}{dx} = -U_H T_H + U_H T_{aH} + \frac{4\eta q^2}{H^3}(4\beta^2 + 2\beta + 1). \tag{3.28a}$$

Both T_{cm} and T_H are functions of position x along the flow channel, and neither is known, so this equation cannot be solved without further information relating T_{cm} and T_H. We have lost this information by integrating Equation 3.19 to obtain an equation for the cup-mixing temperature. If there is wall cooling, we expect that T_H will be the lowest temperature in the system at each value of x, so we expect $T_H(x) < T_{cm}(x)$, $0 < x \le L$. We could make a crude estimate that $T_H(x) \cong T_{cm}(x)$, but it is probably better to write $T_H(x) = \Lambda T_{cm}(x)$, $\Lambda \le 1$ and to take Λ as a constant over the length of the flow channel. (This is also a very crude assumption, but we shall see that it will suffice for our purposes here.) We can then write Equation 3.28a as

$$\rho c_p q \frac{dT_{cm}}{dx} + \Lambda U_H T_{cm} = U_H T_{aH} + \frac{4\eta q^2}{H^3}(4\beta^2 + 2\beta + 1). \tag{3.28b}$$

This is a linear, first-order, ordinary differential equation with constant coefficients and a constant forcing term, for which we can immediately write the solution (after a bit of algebra) as

$$T_{cm} = T_i \exp(-\Lambda x/\chi L) + \frac{1}{\Lambda} [T_{aH} + \chi (T_{ad} - T_i)] [1 - \exp(-\Lambda x/\chi L)], \quad (3.29a)$$

where

$$\chi = \rho c_p q / L U_H \qquad (3.29b)$$

and T_{ad} is the adiabatic solution given in Equation 3.27. The striking result here is that the cup-mixing temperature cannot become fully developed (i.e., independent of x) even with wall cooling unless $\chi \ll 1$ [$\exp(-\Lambda x/\chi L) \to 0$ for $x/L < 1$], which is unlikely. This conclusion is not affected by uncertainties in Λ, which is unlikely to be outside the range $0.5 \le \Lambda < 1$ for absolute temperatures. (Surprisingly, there are very limited data available to estimate U_H. Hay and co-workers report values in the range 3–90 [and one 240] in SI units for seven experiments with polyethylene and polystyrene in slit dies. The limited data do not correlate well on a traditional dimensionless plot, making extrapolation to other conditions difficult. $\chi = 8.4/U$ in SI units for the parameters used in Examples 3.1 and 3.2. The best estimate of U from the dimensionless plot of Hay and co-workers for these conditions is of order 0.1; in that case $\chi \gg 1$. This value seems unreasonable, however, because of the large multiplier of the adiabatic temperature rise and the concomitant low wall temperature required to ensure that the temperature rise with cooling cannot exceed the adiabatic temperature rise. This is an area in which good data are critically needed.)

Finally, it is useful to have an estimate of the fully developed temperature distribution, even though we know that it cannot be reached, since the maximum fully developed temperature will establish a bound on the temperature anywhere in the flow channel. We again return to Equation 3.19, and we now assume that T is independent of x; $T = T(y)$ and $\partial T/\partial x = 0$. The equation then becomes

$$\frac{d^2 T}{dy^2} = -\frac{\eta}{\kappa} \left(\frac{dv_x}{dy} \right)^2 = -\frac{\eta}{\kappa} (\alpha y - \gamma)^2 = -\frac{\eta}{\kappa} (\alpha^2 y^2 - 2\alpha\gamma y + \gamma^2). \qquad (3.30)$$

The function is a quartic if the second derivative is a quadratic; it is straightforward to integrate Equation 3.30 twice to obtain

$$\frac{dT}{dy} = -\frac{\eta}{\kappa} \left(\frac{\alpha^2 y^3}{3} - \alpha\gamma y^2 + \gamma^2 y \right) + C_1, \qquad (3.31a)$$

$$T = -\frac{\eta}{\kappa} \left(\frac{\alpha^2 y^4}{12} - \frac{\alpha\gamma y^3}{3} + \frac{\gamma^2 y^2}{2} \right) + C_1 y + C_2. \qquad (3.31b)$$

Equation 3.31a contains some interesting physical insight. If we specify the flux $-\kappa dT/dy$ at one surface, it is fixed at the other. This is reasonable, since by requiring that the flow be fully developed we have lost a degree of freedom.

We continue to assume that the moving surface at $y = 0$ is adiabatic, so $dT/dy = 0$ at $y = 0$. It then follows from Equation 3.31a that $C_1 = 0$. Setting $y = 0$

in Equation 3.31b, we find that $C_2 = T_o$, the polymer temperature at $y = 0$; this is the maximum temperature in the system, and it is unknown. We can then write

$$T = T_o - \frac{\eta}{\kappa}\left(\frac{\alpha^2 y^4}{12} - \frac{\alpha\gamma y^3}{3} + \frac{\gamma^2 y^2}{2}\right). \tag{3.32}$$

We now impose the heat transfer boundary condition at $y = H$:

$$-\kappa \frac{dT}{dy}\bigg|_{y=H} = \eta\left(\frac{\alpha^2 H^3}{3} - \alpha\gamma H^2 + \gamma^2 H\right) = U_H(T_H - T_{aH})$$

$$= U_H\left[T_o - \frac{\eta}{\kappa}\left(\frac{\alpha^2 H^4}{12} - \frac{\alpha\gamma H^3}{3} + \frac{\gamma^2 H^2}{2}\right) - T_{aH}\right]. \tag{3.33}$$

With some considerable rearrangement, we can solve for T_o as follows:

$$T_o = T_{aH} + \chi\,(T_{ad} - T_i)\left[1 + \frac{6\beta^2 + 4\beta + 1}{2\,(4\beta^2 + 2\beta + 1)}\,Bi\right]. \tag{3.34a}$$

The *Biot number, Bi*, is defined

$$Bi = HU_H/\kappa. \tag{3.34b}$$

(Hay and co-workers refer to this quantity as the *Nusselt number*, but the Nusselt number is properly defined using the thermal conductivity of the ambient medium, not of the polymer.) *Bi* is a measure of the efficiency of heat transfer to the surroundings (U_H) relative to heat transfer by conduction through the polymer (κ/H). The thermal diffusivity ($\kappa/\rho c_p$) of most polymers is about $10^{-7}\,\text{m}^2/\text{s}$, giving a value of κ of about 0.27 W/(m K) for polyethylenes. The Biot number would thus be of order 10^{-2} for $U \sim 0.1$, but of order unity for $U \sim 10$. The latter value seems more realistic. Note that the coefficient of *Bi* is insensitive to β, varying only between 0.5 and about unity and approaching the asymptotic value of 0.75 for $\beta \sim 1$.

3.3.3 Final Comments on the Elementary Extruder

The simple extruder design analyzed here would not be implemented in practice because of obvious mechanical problems, but, as we shall see subsequently, it is sufficiently close to the description of a true single-screw extruder that the calculations done here are all relevant. There are three weaknesses in the analysis. First, we have considered only a Newtonian fluid, while most real polymers have highly shear-dependent viscosities. Second, our heat transfer analysis is inadequate, both because we have considered temperature- and pressure-independent physical properties and because we have been able to obtain explicit solutions only for certain limiting cases. Finally, we have not dealt with the flow in the neighborhood of the transition from the extruder channel to the die. All of these restrictions can be relaxed, as we shall see, but to do so for the latter two generally requires the use of numerical algorithms to solve the full equation set. We shall address this topic in Chapter 8.

3.4 Plane Channel Revisited: Power-Law Fluid

One important limitation in the preceding analysis is the restriction to a Newtonian fluid. This is easily resolved, but at the expense of considerable algebraic complexity, and we simply indicate the approach here without going into the details of the solution. For simplicity we will assume that the fluid viscosity is described by the power-law Equation 2.40a.

Our starting point is again Equation 3.7, which was derived for any fluid with the single assumption that the velocity is of the form $v_y = v_z = 0$ and that v_x is a function only of y. With this assumption, the only nonzero term in the function II_D (Table 2.6) is the one involving dv_x/dy, so we obtain

$$\eta = K \left[\left(\frac{dv_x}{dy} \right)^2 \right]^{\frac{n-1}{2}} = K \left| \frac{dv_x}{dy} \right|^{n-1}. \tag{3.35}$$

The absolute value sign is required to ensure the positivity of η. Equation 3.35 is equivalent to Equation 2.38. Equation 3.7 then becomes

$$\tau_{yx} = K \left| \frac{dv_x}{dy} \right|^{n-1} \frac{dv_x}{dy} = \frac{\partial \mathcal{P}}{\partial x} y + C_1. \tag{3.36}$$

Equation 3.36 is still straightforward to solve by quadrature (direct integration), but there is a complication. dv_x/dy can be positive or negative, depending on the design parameters and on the position between the two walls; compare Figure 3.3 for the special case $n = 1$, which we may assume to be qualitatively representative of all values of n. For $dv_x/dy > 0$, $|dv_x/dy| = dv_x/dy$, while for $dv_x/dy < 0$, $|dv_x/dy| = -dv_x/dy$. Thus, we must write

$$\tau_{yx} = K \left(\frac{dv_x}{dy} \right)^n = \frac{\partial \mathcal{P}}{\partial x} y + C_1 \geq 0, \qquad \frac{dv_x}{dy} > 0, \tag{3.37a}$$

$$\tau_{yx} = -K \left(-\frac{dv_x}{dy} \right)^n = \frac{\partial \mathcal{P}}{\partial x} y + C_1 \leq 0, \qquad \frac{dv_x}{dy} < 0. \tag{3.37b}$$

The simplest case to analyze is the one for which $\partial \mathcal{P}/\partial x > 0$ but $dv_x/dy \leq 0$ over the entire gap width, corresponding to a pressure-driven backflow that is smaller in magnitude at each value of y than the drag flow caused by the moving plate. We can then write Equation 3.37b as

$$\frac{dv_x}{dy} = -\frac{1}{K^{1/n}} \left[-\frac{\partial \mathcal{P}}{\partial x} y - C_1 \right]^{1/n} = -\left(\frac{1}{K} \frac{\partial \mathcal{P}}{\partial x} \right)^{1/n} (C_2 - y)^{1/n}, \tag{3.38}$$

where $C_2 \geq H$ is a new constant defined as $-C_1/(\partial \mathcal{P}/\partial x)$. C_2 is positive, since we know from Equation 3.37b that $C_1 \leq 0$. Note the care with which it has been necessary to keep track of positive quantities in order to avoid the possibility of seeking fractional powers of negative numbers.

Equation 3.38 is readily integrated, most easily by changing the independent variable from y to $C_2 - y$. The result is

$$v_x(y) = \frac{n}{n+1}\left(\frac{1}{K}\frac{\partial \mathcal{P}}{\partial x}\right)^{1/n}(C_2 - y)^{(n+1)/n} + C_3. \tag{3.39}$$

The constants C_2 and C_3 are obtained from the boundary conditions:

$$\text{at } y = 0, v_x = V = \frac{n}{n+1}\left(\frac{1}{K}\frac{\partial \mathcal{P}}{\partial x}\right)^{1/n}C_2^{(n+1)/n} + C_3, \tag{3.40a}$$

$$\text{at } y = H, v_x = 0 = \frac{n}{n+1}\left(\frac{1}{K}\frac{\partial \mathcal{P}}{\partial x}\right)^{1/n}(C_2 - H)^{(n+1)/n} + C_3. \tag{3.40b}$$

C_2 is thus the solution of the equation

$$C_2^{(n+1)/n} - (C_2 - H)^{(n+1)/n} = \frac{n+1}{n}\left(\frac{1}{K}\frac{\partial \mathcal{P}}{\partial x}\right)^{-1/n}V, \tag{3.41}$$

while C_3 is obtained from Equation 3.40b, and the velocity profile is

$$v_x(y) = \frac{n}{n+1}\left(\frac{1}{K}\frac{\partial \mathcal{P}}{\partial x}\right)^{1/n}\left[(C_2 - y)^{(n+1)/n} - (C_2 - H)^{(n+1)/n}\right]. \tag{3.42}$$

Equation 3.41 is linear in C_2 for the special case $n = 1$, and it is readily established in that case that Equation 3.42 reduces to Equation 3.11. It is a polynomial equation of order m for integer values $m = 1/n$, but except for a few special cases numerical solution is required. Note that the requirement $v_x \geq 0$ for all y does not permit us to go to the limit $V \to 0$.

We could continue this analysis to parallel the full development for $n = 1$, but to do so would not be informative. The cases of real interest, of course, are those for which $\partial \mathcal{P}/\partial y$ is sufficiently large to cause a negative velocity over some part of the channel, corresponding to profiles like these in Figure 3.3(c) and (d) for $n = 1$. The analysis is carried out exactly as above, but now it is necessary to divide the channel into two regions, one where $dv_x/dy < 0$ and one where $dv_x/dy > 0$. We solve Equation 3.37b in the former and Equation 3.37a in the latter. The (unknown) value of y where the two solutions must match is given by $dv_x/dy = 0$. It is readily established that the no-slip boundary condition at $y = 0$ and H and the condition that the velocity be continuous at the point dividing the two regions (where $dv_x/dy = 0$) are sufficient to evaluate all constants of integration.

3.5 Single-Screw Extruder

3.5.1 Geometry and Kinematics

The analysis in the preceding sections is readily extended to describe flow in the melt zone of a single-screw extruder. A section of the extruder is shown schematically in Figure 3.4. The screw has a radius R, and the spacing between the screw and the barrel is H. The flights are at an angle θ to the screw axis, and the spacing between the flights measured orthogonal to the flight surface is W (that is, the

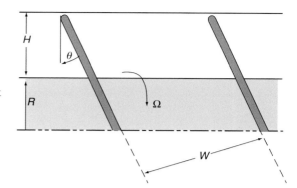

Figure 3.4. Schematic of a section of the melt
zone of a single-screw extruder (not to scale).

distance between flights along the screw axis is $W \cos \theta$). We assume $R \gg H$ and
$W \gg H$. We also assume that there is no "leakage" flow through the space between
the flights and the barrel. The screw turns with an angular velocity Ω. We are assuming here that H is a constant throughout the melt zone, which may not be true, but
this assumption is easily relaxed using the methodology developed in the next chapter. The assumption $W \gg H$ may be a poor one, and it, too, can be relaxed, but only
by turning to a numerical solution.

The critical assumption enabling us to use the results in the preceding section
is $H \ll R$. In that case, observers in the gap between the screw and the barrel
would be unaware of the curvature and would feel as though they were in a plane
channel (just as humans are unaware of the local curvature of the surface of the
earth); the further assumption $W \gg H$ places the side walls far away, giving the
channel the appearance of the gap between two infinite plates. The "unwrapped"
screw is shown schematically in Figure 3.5. The x axis is aligned with the helical
channel. The relative linear velocity between the plates is $R\Omega$, with components
$V_x = R\Omega \cos \theta$ and $V_z = R\Omega \sin \theta$.

The approach is identical to that in Section 3.2, except that now, because the
direction of rotation of the screw is at an angle θ to the direction of the flow channel,
we have both x and z components of velocity at the surface $y = 0$, hence, finite
v_x and v_z. We assume $v_y = 0$ and $v_x = v_x(y)$, $v_z = v_z(y)$; the continuity equation
$\partial v_x / \partial x + \partial v_y / \partial y + \partial v_z / \partial z = 0$ is thus automatically satisfied. The same arguments
used in Section 3.2 permit us to write the components of the creeping flow equations
as

$$x: 0 = -\frac{\partial \mathcal{P}}{\partial x} + \frac{\partial \tau_{yx}}{\partial y}, \tag{3.43a}$$

$$y: 0 = -\frac{\partial \mathcal{P}}{\partial y} + \frac{\partial \tau_{yy}}{\partial y}, \tag{3.43b}$$

$$z: 0 = -\frac{\partial \mathcal{P}}{\partial z} + \frac{\partial \tau_{yz}}{\partial y}. \tag{3.43c}$$

It then follows that

$$\mathcal{P} = \tau_{yy} + \text{function of } (x, z), \tag{3.44}$$

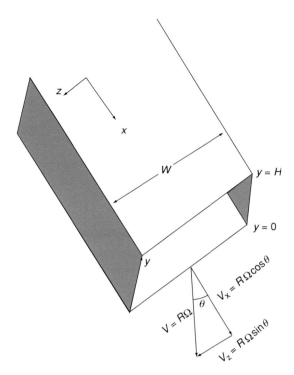

Figure 3.5. Schematic of an unwrapped section of the melt zone of a single-screw extruder.

where τ_{yy} cannot depend on y. Hence,

$$\frac{\partial \mathcal{P}}{\partial x} = \frac{d\tau_{yx}}{dy} = \text{constant}, \quad \frac{\partial \mathcal{P}}{\partial z} = \frac{d\tau_{yz}}{dy} = \text{constant}, \quad (3.45\text{a,b})$$

where we conclude that the terms in Equations 3.45a and 3.45b, respectively, must equal constants because the left-hand sides are independent of y and the right-hand sides depend on y. Finally, then, in analogy to Equation 3.7, we write

$$\tau_{yx} = \frac{\partial \mathcal{P}}{\partial x} y + C_1^x \quad \tau_{yz} = \frac{\partial \mathcal{P}}{\partial z} y + C_1^z, \quad (3.46\text{a,b})$$

where C_1^x and C_1^z are constants.

3.5.2 Newtonian Fluid

For the Newtonian fluid we write

$$\eta \frac{dv_x}{dy} = \frac{\partial \mathcal{P}}{\partial x} y + C_1^x, \quad (3.47\text{a})$$

$$\eta \frac{dv_z}{dy} = \frac{\partial \mathcal{P}}{\partial z} y + C_1^z. \quad (3.47\text{b})$$

These equations are completely uncoupled, so we may immediately apply the solution given by Equations 3.11 and 3.12, simply replacing V by $R\Omega \cos\theta$ and $R\Omega \sin\theta$

for the x and z components, respectively:

$$v_x(y) = \frac{H^2}{2\eta} \frac{\partial \mathcal{P}}{\partial x} \frac{y}{H} \left(\frac{y}{H} - 1\right) + R\Omega \cos\theta \left(1 - \frac{y}{H}\right), \qquad (3.48a)$$

$$v_z(y) = \frac{H^2}{2\eta} \frac{\partial \mathcal{P}}{\partial z} \frac{y}{H} \left(\frac{y}{H} - 1\right) + R\Omega \sin\theta \left(1 - \frac{y}{H}\right), \qquad (3.48b)$$

$$q_x = \int_0^H v_x dy = -\frac{H^3}{12\eta} \frac{\partial \mathcal{P}}{\partial x} + \frac{R\Omega H \cos\theta}{2}, \qquad (3.49a)$$

$$q_z = \int_0^H v_z dy = -\frac{H^3}{12\eta} \frac{\partial \mathcal{P}}{\partial z} + \frac{R\Omega H \sin\theta}{2}. \qquad (3.49b)$$

Now, q_x is the flow rate in the channel direction, Q, divided by the width, W, and it is the quantity of processing interest. If we interpret L as the length of the melt zone measured along the helical channel, we can then simply repeat the analysis in Section 3.3.1 leading up to Equation 3.16 without change, except to substitute Q/W for q_x and $R\Omega \cos\theta$ for V, to obtain

$$\Omega = \frac{2Q(1+\beta)}{WHR\cos\theta}, \qquad (3.50)$$

$$\frac{v_x}{R\Omega \cos\theta} = \frac{3\beta}{(1+\beta)}\xi(\xi - 1) + (1 - \xi), \qquad (3.51)$$

where $\beta = (H/h)^3 l/L$, as defined previously, and $\xi = y/H$. (β will have a different definition if the die cross section has a shape different from a slit. For a circular die of radius R and length l, for example, $\beta = 2WH^3l/3\pi R^4 L$.) If we assume no leakage across the flights, there can be no *net* flow in the z direction; that is, $q_z = 0$ and

$$\frac{\partial \mathcal{P}}{\partial z} = \frac{6\eta R\Omega \sin\theta}{H^2}. \qquad (3.52a)$$

Hence, Equation 3.48b becomes

$$\frac{v_z}{R\Omega \sin\theta} = 3\xi(\xi - 1) + (1 - \xi). \qquad (3.52b)$$

This is simply Equation 3.16a with $\beta \to \infty$ (pressure flow = drag flow, Figure 3.3d).

The power consumption calculation in Equations 3.17 and 3.18, with q replaced by $q_x = Q/W$, gives only the portion of the power associated with the flow along the the channel. To this must be added the power consumption associated with cross-channel flow, $\tau_{yz}|_{y=0} LWR\Omega\sin\theta$. The calculation of the additional term is straight-forward, and with some algebra we obtain the result

$$\wp = \frac{4\eta LQ^2}{WH^3}(1+\beta)(1+4\beta)\left[1 + 4\frac{1+\beta}{1+4\beta}\tan^2\theta\right]. \qquad (3.53)$$

The $\tan^2\theta$ correction term will generally be small compared to unity.

3.5.3 Temperature Development

Development of the temperature profile in the unwrapped model of the single-screw extruder is essentially unchanged from that given in Section 3.3.2. The dissipation term in the energy equation contains an additional contribution $\eta(dv_x/dy)^2$, but this is again simply a quadratic in y and leads to a correction approximately equal to $1 + 4\tan^2\theta$, which can easily be incorporated. Equation 3.19 is replaced by

$$\rho c_p \left(v_x \frac{\partial T}{\partial x} + v_z \frac{\partial T}{\partial z} \right) = \kappa \frac{\partial^2 T}{\partial y^2} + \eta \left[\left(\frac{dv_x}{dy} \right)^2 + \left(\frac{dv_z}{dy} \right)^2 \right]. \tag{3.54}$$

As before, we assume that conduction is negligible relative to convection in directions in which there is flow.

We cannot expect $\partial T/\partial z$ to vanish, even away from the flights (where the flow is far more complex than the one we are considering here, but where the overall contribution is small because of the assumption $H \ll W$), because fluid is being convected downstream even as it flows transversely across the channel. We can get a sense of what is happening by defining an average temperature with respect to z at each x and y as

$$\overline{T}(x, y) = \frac{1}{W} \int_0^W T(x, y, z)dz. \tag{3.55}$$

Integrating each term in Equation 3.54 with respect to z and dividing by W then gives

$$\rho c_p v_x \frac{\partial \overline{T}}{\partial x} + \rho c_p v_z [T(x, y; z = W) - T(x, y; z = 0)]$$
$$= \kappa \frac{\partial^2 \overline{T}}{\partial y^2} + \eta \left[\left(\frac{dv_x}{dy} \right)^2 + \left(\frac{dv_z}{dy} \right)^2 \right]. \tag{3.56}$$

(The limits of integration in Equation 3.55 are a bit delicate. Clearly, v_z vanishes at $z = 0$ and $z = W$, the locations of the flights, but our assumed flow pattern is also in error there. Thus, we should think of these limits as being "near" $z = 0$ and $z = W$, but still in a region where the kinematic assumptions apply.) The second term on the left represents a contribution to temperature profile development from the transverse flow. We have no a priori way of estimating its magnitude, although we expect it to be small, and in general we will estimate thermal effects by setting $T(x, y; z = W) = T(x, y; z = 0)$, in which case the analysis is identical to that in Section 3.3.2, except for multiplication of the dissipation term by a factor of approximately $1 + 4\tan^2\theta$.

3.5.4 Power-Law Fluid

For a power-law fluid,

$$\eta = K \left| \left(\frac{dv_x}{dy} \right)^2 + \left(\frac{dv_z}{dy} \right)^2 \right|^{\frac{n-1}{2}}, \tag{3.57}$$

Equations 3.46a and 3.46b are now coupled, and an analytical solution is not possible. Based on our analysis of the dissipation term for a Newtonian fluid, we could anticipate that we will often have $(dv_x/dy)^2 \gg (dv_z/dy)^2$ and neglect the second term, in which case the equation for v_x can be solved independently of the equation for v_z. We would not obtain any new insight into the physical processes, however, so we will not pursue this approach. Modern computational tools, which we will discuss in Chapter 8, provide a more useful approach to the solution of problems with complex rheology once we have a good understanding of the basic phenomena from appropriate limiting cases, including the Newtonian liquid.

BIBLIOGRAPHICAL NOTES

The material in this chapter is discussed from other perspectives in

Middleman, S., *Fundamentals of Polymer Processing*, McGraw-Hill, New York, 1977.
Pearson, J. R. A., *Mechanics of Polymer Processing*, Elsevier Applied Science, London, 1985.
Tadmor, Z., and C. G. Gogos, *Principles of Polymer Processing,* 2nd ed., Wiley InterScience, New York, 2006.

Limited data on heat transfer in slits may be found in

Hay, G., M. E. Mackay, K. M. Awati, and Y. Park, *J. Rheol.*, **43**, 1099 (1999).

4 Temperature and Pressure Effects in Flow

4.1 Introduction

The analysis of extrusion in the preceding chapter was based on the assumption that the temperature and pressure dependence of physical properties, especially the viscosity, could be neglected. This assumption simplifies the analysis, especially in the case of temperature dependence, because it introduces an uncoupling between the fluid mechanics and the heat transfer. The assumption is dangerous if not used with care, however, as we shall demonstrate in this chapter.

4.2 Pressure-Dependent Viscosity and "Choking"

4.2.1 Isothermal Flow

The viscosity of organic liquids depends on both temperature and pressure; molecular motion becomes more difficult as free volume is reduced, and the viscosity increases. To a first approximation, the viscosity of polymer melts can be written[*]

$$\eta = \eta^o e^{-\alpha(T-T_o)} e^{\beta p}. \tag{4.1}$$

η^o is the viscosity at atmospheric pressure ($p = 0$) and the reference temperature T_o; η^o may depend on the shear rate. β is typically $1\text{–}5 \times 10^{-8}$ Pa^{-1}, while α is typically $1\text{–}8 \times 10^{-2}$ K^{-1}. Thus, temperature differences of 10 degrees can have a significant effect on the viscosity, and we expect the pressure dependence to become important at a pressure of about 5×10^6 Pa (50 atm), which can be reached in extrusion and is routinely seen in injection molding. The density change at these elevated pressures is small, so compressibility is rarely important, and it usually suffices to retain the incompressible form of the continuity equation, even when accounting for the pressure dependence of the viscosity. For illustrative purposes, it suffices to consider only the case of die flow, where the wall velocity $V = 0$ and the exit

---

[*] $\exp[-\alpha(T - T_o)]$ is an approximation to an Arrhenius relation. $\exp[-(E/T_o - E/T)] = \exp[-E(T - T_o)/T T_o] \sim \exp[-E(T - T_o)/T_o^2]$. Thus, $\alpha = E/T_o^2$, where E is the activation energy in units of absolute temperature.

pressure is zero. We also consider only the case of a Newtonian fluid, although the generalization to a shear rate-dependent viscosity is straightforward.

We start by considering an isothermal flow, where $T = T_o$ everywhere. As before, we assume that the flow is rectilinear, so the only nonzero velocity component is v_x, and v_x is a function only of y. The x component of the momentum equation then becomes

$$\frac{\partial p}{\partial x} = \frac{\partial}{\partial x} \tau_{yx} = \frac{\partial}{\partial y} \eta^o e^{\beta p} \frac{dv_x}{dy}. \tag{4.2}$$

To proceed further, we need to assume that $p = p(x)$; that is, the pressure is independent of the transverse coordinate. This is at best an approximation because it readily follows from the y component of the momentum equation that p *cannot* be a function of x alone. The transverse dependence of p can be shown to be very small, however, and in fact an exact solution without the approximation $p = p(x)$ is available for flow in a cylindrical cross section. We will therefore proceed by taking p to be independent of y. Equation 4.2 can then be written

$$e^{-\beta p} \frac{dp}{dx} = \frac{d}{dy} \eta^o \frac{dv_x}{dy}. \tag{4.3}$$

The left-hand side of Equation 4.3 is a function only of x, while the right-hand side is a function of y, so both must be constant. The right-hand side can be integrated for any viscosity function $\eta^o(\dot{\gamma})$; for a Newtonian fluid we simply obtain the parabolic velocity distribution,

$$v_x = \frac{6q}{H^3} y(H - y). \tag{4.4}$$

It then follows that

$$e^{-\beta p} \frac{dp}{dx} = -\frac{1}{\beta} \frac{d}{dx} e^{-\beta p} = -\frac{12\eta^o q}{H^3} \tag{4.5}$$

and, with $p = 0$ at $x = L$,

$$p = -\frac{1}{\beta} \ln \left[1 - \frac{12\eta^o \beta q}{H^3} (L - x) \right]. \tag{4.6}$$

The overall pressure drop across the die is then

$$|\Delta p| = -\frac{1}{\beta} \ln \left[1 - \frac{12\eta^o \beta q L}{H^3} \right] \tag{4.7a}$$

or

$$|\Delta p| = -\frac{1}{\beta} \ln \left[1 - \beta |\Delta p^o| \right], \tag{4.7b}$$

where $|\Delta p^o|$ is the pressure drop for $\beta = 0$ (i.e., no pressure dependence of viscosity). It is easily shown that Equation 4.7b is valid for any viscosity dependence on shear rate with an appropriate interpretation of $|\Delta p^o|$, and not just for a Newtonian fluid.

The striking thing about Equation 4.7 is that the pressure will begin to increase rapidly with flow rate, and will become infinite for a finite flow rate. Hence, there is a maximum flow rate that is possible, regardless of the pressure drop. This phenomenon is known as *choking*, and it is analogous to a superficially similar phenomenon in compressible gas dynamics.

It is of course possible to redo the simple extruder analysis with a pressure-dependent viscosity. Equation 4.7b defines the die characteristic equation. Solution of the momentum equation with a moving wall then leads directly to Equation 3.15b for the wall velocity as a function of the flow rate and system geometry, even when the viscosity is pressure dependent.

4.2.2 Viscous Dissipation

Choking flow is rarely, if ever, encountered in practice, although pressure increases because of the pressure dependence of the viscosity do occur. The reason for the absence of choking is the effect of viscous dissipation, which by itself causes the viscosity to *decrease*. The problem of die flow with a viscosity that is dependent on both temperature and pressure can be attacked fully only by numerical methods, but substantial insight can be obtained from a rather simple approximate analytical solution.

We start with the energy and momentum equations in their most general form (except for conduction in the flow direction) for one-dimensional flow:

$$\rho c_p v_x \frac{\partial T}{\partial x} = \kappa \frac{\partial^2 T}{\partial y^2} + \tau_{yx} \frac{\partial v_x}{\partial y}, \tag{4.8a}$$

$$\frac{dp}{dx} = \frac{d\tau_{yx}}{dy}. \tag{4.8b}$$

The solution to Equation 4.8b, with the assumption that $p = p(x)$, is

$$\tau_{yx} = \frac{dp}{dx}(y - C), \tag{4.9}$$

where C is a constant. We now integrate Equation 4.8a from $y = 0$ to $y = H$. It is straightforward to show, using Equation 4.9, that

$$\int_0^H \tau_{yx} \frac{dv_x}{dy} dy = -q \frac{dp}{dx}. \tag{4.10}$$

Hence, the integrated form of Equation 4.8a is (cf. Equation 3.21)

$$\rho c_p q \frac{dT_{cm}}{dx} = \kappa \frac{\partial T}{\partial y}\Big|_{y=0}^{y=H} - q \frac{dp}{dx}. \tag{4.11}$$

The first term on the right is expressed in terms of a heat transfer coefficient and a temperature driving force. We limit ourselves here to adiabatic flow, in which case the first term on the right vanishes and

$$\rho c_p \frac{dT_{cm}}{dx} = -\frac{dp}{dx}. \tag{4.12}$$

For constant ρc_p we then obtain

$$\rho c_p \left(T_{cm} - T_i\right) = -\left(p - p_i\right) = |\Delta p| - p, \tag{4.13}$$

where the subscript i denotes inlet conditions and we have taken the exit pressure to be zero.

We now assume that we can replace the temperature in the viscosity with the cup-mixing temperature and write

$$\eta = \eta^o e^{-\alpha(T_{cm} - T_i)} e^{\beta p}, \tag{4.14}$$

where we have chosen the reference temperature T_o to be the inlet temperature. The approximation that the temperature over the entire cross section equals the cup-mixing temperature is crude at best, but it will suffice for the insight that we seek. In that case, the viscosity can be written

$$\eta = \eta^o e^{-\alpha|\Delta p|/\rho c_p} e^{\left(\beta + \frac{\alpha}{\rho c_p}\right)p}. \tag{4.15}$$

$\rho c_p \beta/\alpha$ is typically of order unity, so the two terms multiplying p in the exponential are of comparable importance.

The viscosity is now an exponential function only of pressure, and, with some change of nomenclature (η^o replaced with $\eta^o \exp(-\alpha|\Delta p|/\rho c_p)$ and β with $\beta + \alpha/\rho c_p$), the problem is identical to the one already solved in the preceding section. Using Equation 4.8b we can then immediately write

$$\left(\beta + \frac{\alpha}{\rho c_p}\right)|\Delta p| = -\ln\left[1 - \left(\beta + \frac{\alpha}{\rho c_p}\right)|\Delta p^o| e^{-\alpha|\Delta p|/\rho c_p}\right] \tag{4.16a}$$

or, equivalently,

$$\left(\beta + \frac{\alpha}{\rho c_p}\right)|\Delta p^o| = e^{\alpha|\Delta p|/\rho c_p} - e^{-\beta|\Delta p|}. \tag{4.16b}$$

Equation 4.16b is an explicit expression for the flow rate in terms of the pressure drop. The second term on the right can be neglected with less than 5% error for $(\beta + \alpha/\rho c_p)|\Delta p| > 3$, which corresponds roughly to pressure drops greater than 10^8 Pa. In that case we obtain an explicit expression for the pressure drop and the adiabatic temperature rise in terms of the flow rate:

$$\alpha\left(T_{cm} - T_i\right) = \frac{\alpha|\Delta p|}{\rho c_p} \approx \ln\left[\left(\beta + \frac{\alpha}{\rho c_p}\right)|\Delta p^o|\right]. \tag{4.17}$$

Hence, choking never occurs when viscous heating is taken into account. The adiabatic pressure drop without accounting for the pressure dependence of viscosity follows from Equation 4.16b by letting $\beta \to 0$:

$$\beta \to 0: \quad |\Delta p| = \frac{\rho c_p}{\alpha} \ln\left(1 + \frac{\alpha|\Delta p^o|}{\rho c_p}\right). \tag{4.18}$$

The same equations hold for flow through a round capillary.

The adiabatic pressure drop with and without taking the pressure dependence of viscosity into account is shown in Figure 4.1, where a value of $\beta = 2.5\alpha/\rho c_p$ was used. These normalized curves depend only on the relative values of α and β. Some

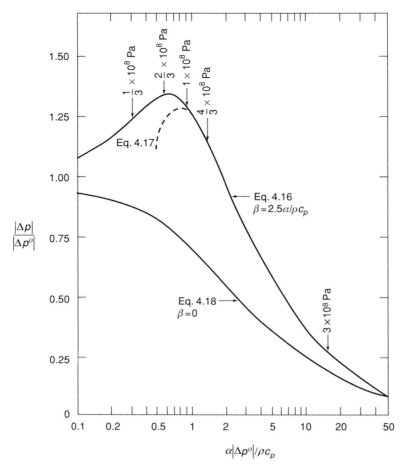

Figure 4.1. Effect of temperature and pressure dependence of viscosity on pressure drop for adiabatic flow through a plane channel or tube. Reprinted from Denn, *Polym. Eng. Sci.*, **21**, 65 (1981).

computed pressure drops are shown for $\beta = 3 \times 10^{-8}\,\mathrm{Pa}^{-1}$, which is typical of polymers of commercial interest. A substantial increase in pressure drop above that for isothermal flow is found over part of the region, which corresponds to the region in which injection molding is practiced.

4.3 Multiplicity and Instability

The coupling of flow and thermal effects with a temperature-dependent viscosity can lead to situations in which more than one solution to the model equations is possible, suggesting that the process can exist in more than one state for a given set of flow conditions. This phenomenon of *multiplicity* is often a surprise when first encountered, but it is a common characteristic of nonlinear systems. In combustion, for example, it is known that there is a finite range of flow conditions for which the system can exist in either of two states: a low-temperature, low-conversion state, and a high-temperature, high-conversion state. (A third intermediate state is also a solution of the combustion equations but is inherently unstable and cannot be observed experimentally.) Laminar (streamline) flow is an exact solution of the Navier-Stokes equations for isothermal tube flow at Reynolds numbers above 2,100, but turbulence is the flow state that is usually observed, indicating the existence of a second solution to the equations; indeed, laminar flow has been observed up to Reynolds numbers of 50,000 in careful vibration-free experiments.

Consider the following scenario for flow of a polymer melt. Hot polymer is fed to a channel with wall cooling at a fixed overall pressure drop. If the flow rate is small, there will be substantial cooling because of the long residence time in the channel, and the viscosity will increase. If the flow is rapid, however, there will be little cooling, the temperature will remain high, and the viscosity will be low. A low viscosity supports a fast flow for a given pressure drop, while a high viscosity supports a slow flow for the same pressure drop. It is possible that there can be conditions where low-flow-rate/high-viscosity and high-flow-rate/low-viscosity solutions both correspond to the same pressure drop. The same scenario can exist with a cold feed if we introduce viscous heating, since the viscous heating at a high flow rate can serve to increase the temperature and obviate the need for a hot feed.

We now consider the quantitative treatment of this problem. For simplicity we will consider the case of a plane channel in which the feed temperature T_i is greater than the ambient temperature T_a, and we will neglect viscous dissipation. The starting point is then Equation 4.11 without the $q\,dp/dx$ term. We write the wall heat transfer rate from the melt to ambient as $U(T_{cm} - T_a)$; this definition of the heat transfer coefficient suffices to illustrate the point that we wish to make, and the results would be essentially the same if we followed the procedure used for the extruder and set the wall temperature to a fixed fraction of T_{cm}. There are two walls in the channel, so Equation 4.10 then becomes

$$\rho c_p q \frac{dT_{cm}}{dx} = -2U(T_{cm} - T_a). \tag{4.19}$$

It is convenient to define dimensionless quantities as follows:

$$\Theta = \alpha(T_{cm} - T_a), \quad B = \alpha(T_i - T_a), \quad \xi = x/L, \quad Gz = \rho c_p q / 2UL.$$
$$\text{(4.20a,b,c,d)}$$

Gz is known as the *Graetz number*; for this problem we can think of it as a dimensionless flow rate. Equation 4.19 then becomes

$$Gz \frac{d\Theta}{d\xi} = -\Theta, \tag{4.21}$$

which has a solution

$$\Theta = Be^{-\xi/Gz}. \tag{4.22}$$

We now turn to the momentum equation,

$$0 = -\frac{dp}{dx} + \frac{\partial}{\partial y}\tau_{yx}, \tag{4.23}$$

which, with one integration, becomes

$$\tau_{yx} = \frac{dp}{dx}\left(y - \frac{H}{2}\right) = \eta^o e^{-\alpha(T-T_a)}\frac{\partial v_x}{\partial y}. \tag{4.24}$$

Here we have taken the reference temperature T_o for the viscosity to be T_a.

The next steps in the development are not intuitive and seem to require that we know where we wish to go. Clearly, we will integrate with respect to y, but the right-hand side of Equation 4.24 presents two difficulties. The obvious one is that we have no detailed information about the temperature profile, so it will be necessary to replace the term $\exp[-\alpha(T - T_a)]$ with a suitable average, which we take to be $\exp(-\Theta)$. The other problem is that direct integration will simply produce $0 = 0$ because of the symmetry of the flow about the center plane and the vanishing of the velocity at the walls. We can get around this difficulty and obtain an equation in terms of q if we multiply both sides of Equation 4.23 by y and integrate from $y = 0$ to $y = H$; one integration by parts, together with Equation 4.24, then yields

$$\frac{dp}{dx} = -\frac{12}{H^3}\eta^o e^{-\Theta}q \tag{4.25}$$

or, integrating with respect to x and making use of the fact that $p(L) = 0$,

$$|\Delta p| = p(0) = \frac{12\eta^o q L}{H^3}\int_0^1 e^{-\Theta}d\xi. \tag{4.26}$$

It is convenient to express this result entirely in terms of the Graetz number, in which case we write

$$\Pi \equiv \frac{\rho c_p H^3 |\Delta p|}{24\eta^o U L^2} = Gz\int_0^1 e^{-B\exp(\xi/Gz)}d\xi. \tag{4.27}$$

The dimensionless pressure drop Π is plotted as a function of Gz in Figure 4.2 for a range of values of the dimensionless temperature difference B. The curve is monotonically increasing for small values of B; that is, the pressure drop increases monotonically with flow rate, as expected. There is an inflection point for $B \sim 3$, however, and for higher values of B there is a maximum and a minimum. Hence, a horizontal line drawn at a pressure drop below the maximum but above the minimum for a given value of B will intersect the curve three times, indicating that three different flow rates will satisfy the momentum and energy equations (with the approximations employed here) for the same pressure drop.

The decreasing branch of the curve, where $d\Pi/dGz < 0$, appears to be unphysical; a small decrease in pressure is predicted to give a small *increase* in flow rate. It is likely that operating points on this branch are unstable and cannot be maintained in practice, but a definitive answer would require an analysis of the full transient equations. If we imagine a sequence of experiments at increasing pressure drop for $B > 3$, we expect an increasing flow rate as we proceed along the first (ascending) branch of the curve. At the maximum there should be a discontinuous jump in flow rate to the third (ascending) branch, and it is the latter branch that is followed for

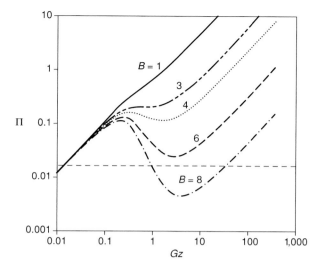

Figure 4.2. Dimensionless pressure drop Π as a function of Graetz number (dimensionless flow rate) for flow through a plane channel with wall cooling. The horizontal line shows three intersections, corresponding to the three values of Gz that satisfy Equation 4.27 at fixed B for a given Π.

still further increases in pressure drop. Conversely, as we decrease the pressure drop from a large value and move down the third branch, the flow rate will decrease continuously until we reach the minimum in the curve. With any further decrease in the pressure drop, there will be a discontinuous jump to the first branch and a large decrease in flow rate. Hence, there will be discontinuities and hysteresis in the flow curve. Steady flow rates between the maximum and the minimum in the curve cannot be achieved under pressure control. (It is possible, of course, that the transition between branches takes place at an intermediate point or that the system oscillates between branches. Pearson and co-workers have done a dynamic stability analysis of this situation, using the methodology described in Chapter 11, and do find, in fact, that the transition occurs before the maximum in the flow curve.)

We have not even speculated on the system response if the flow rate, rather than the pressure drop, is controlled. It is possible that steady operation could be achieved on the descending branch in that case, since the pressure drop is a single-valued function of Gz; that is, there is only one pressure drop for each flow rate.

Viscous heating introduces an additional dimensionless group, and the analysis of the energy equation is slightly less straightforward, but the qualitative behavior is the same. Now, however, a multivalued flow curve can be obtained even for negative values of B, where the feed temperature is less than that of the surroundings.

Finally, we note that the wall cooling is essential for multiplicity to occur for this approximate set of equations. As we saw in the preceding section, the pressure drop is a monotonically increasing function of flow rate for adiabatic flow.

4.4 Concluding Remarks

These two examples were selected to illustrate the interesting behavior that can occur in highly coupled systems and to serve as a caution about the limitations and possible dangers of models that oversimplify the physics by uncoupling phenomena.

Heat transfer, in particular, can be the overriding transport process in many polymer processing applications, and the strong temperature dependence of the viscosity of polymer melts needs to be kept in mind. This, rather than the specific applications considered here, is the major lesson to be learned from this chapter.

BIBLIOGRAPHICAL NOTES

The treatment of the pressure-dependent viscosity follows an analysis for a cylindrical cross section in the following, where an exact solution with a parabolic velocity profile for the isothermal case can be found:

Denn, M. M., *Polym. Eng. Sci.*, **21**, 65 (1981).

Renardy has shown that an exact solution with rectilinear flow exists only for an exponential pressure dependence in a cylindrical cross section and only for a linear pressure dependence in a plane channel:

Renardy, M., *J. Non-Newtonian Fluid Mech.*, **114**, 229 (2003).

There are more recent treatments, which include heat transfer data, in the context of capillary rheometry in

Hay, G., M. E. Mackay, K. M. Awati, and Y. Park, *J. Rheol.*, **43**, 1099 (1999).
Laun, H. M., *Rheolog. Acta*, **42**, 295 (2003).
Laun, H. M., *Rheolog. Acta*, **43**, 509 (2004).

The analysis of developing flow and multiplicities, including non-Newtonian viscosity and viscous dissipation, is treated in a series of articles by Pearson and coworkers:

Pearson, J. R. A., Y. T. Shah, and E. S. A. Viera, *Chem. Eng. Sci.*, **28**, 2079 (1973).
Shah, Y. T., and J. R. A. Pearson, *Chem. Eng. Sci.*, **29**, 737, 1435 (1974).

The work is summarized in

Pearson, J. R. A., *Mechanics of Polymer Processing*, Elsevier Applied Science, London, 1985, pp. 589ff.

The description of wall heat transfer is slightly different from what is used here, but this is an unimportant detail and the results are essentially equivalent. Pearson takes the averaged equations as starting points, and the approximations entailed are not obvious. There are transcription errors in Pearson's book, so the original articles, which also contain more detail, should be consulted.

Fully developed shear flow with a temperature-dependent viscosity and viscous dissipation can develop multiplicities in the shear stress (hence, the pressure drop) for a given shear rate. An article by Sukanek and Laurence demonstrates the existence of a double-valued shear stress experimentally and is a good place to start:

Sukanek, P. C., and R. L. Laurence, *AIChE J.*, **20**, 474 (1974).

The topic is covered in Pearson's book and in a review:

Rauwendaal C. J., and J. F. Ingen Housz, *Int. Polym. Proc.*, **III**, 123 (1988).

A more recent article, with analysis and experiments, is

Skul'skiy, O. I., Y. V. Slanov, and N. V. Shakirov, *J. Non-Newtonian Fluid Mech.*, **81**, 17 (1999).

5 The Thin Gap Approximation

5.1 Introduction

The preceding chapters addressed flows with a single velocity component that is parallel to the conduit walls. Most confined polymer processing operations are characterized by flows in thin gaps, but in many cases the walls are not parallel, so there must be more than one component of velocity. It is often the case, however, that the gap between the confining surfaces changes slowly in the direction of mean flow, a situation we call *nearly parallel*. Such flows can be treated analytically, and we can gain considerable insight into process performance and design. We will illustrate the approach in this chapter with an application of polymer coating of a sheet, but the methodology applies equally well to calendaring, extrusion, and compression and injection molding.

The analysis of nearly parallel flows originated in the study of problems of lubrication, and the approach is often called the *lubrication approximation*. The terminology is unfortunate from our perspective, given that this approach is at the heart of all analytical treatments of polymer processing operations – we would prefer that it be called the *polymer processing approximation* – but the historical name is well established. The major figure in the analysis of lubrication flows was Osborne Reynolds, and one widely used form of the resulting equations is often called the *Reynolds lubrication equation*.

5.2 Basic Equations, Newtonian Liquid

We restrict ourselves to two-dimensional flows, where all changes occur in the xy plane and there is no flow in the "neutral" z direction. We found in Section 3.2 that the following equations apply for parallel flow of a Newtonian fluid with a moving surface at $y = 0$ and a stationary surface at $y = H$:

$$v_x(y) = \frac{H^2}{2\eta} \frac{\partial \mathcal{P}}{\partial x} \frac{y}{H} \left(\frac{y}{H} - 1\right) + V \left(1 - \frac{y}{H}\right), \tag{5.1}$$

Figure 5.1. Schematic of a nearly parallel flow.

$$q = -\frac{H^3}{12\eta}\frac{\partial \mathcal{P}}{\partial x} + \frac{VH}{2}. \qquad (5.2)$$

Now consider flow in the geometry shown in Figure 5.1, where the surface at $y = 0$ moves with velocity V. The stationary surface is defined by the line $y = H(x)$, and the local angle α relative to the surface at $y = 0$ is given by $\alpha = dH/dx$. If α is small, the fluid *locally* experiences an environment that differs from the flow between parallel walls by only a small amount (i.e., with a correction to parallel flow that should be of order α). Thus, for small dH/dx we expect the flow field at each value of x to be closely approximated by Equation 5.1, with H evaluated at the appropriate value $H(x)$. We therefore write

$$v_x(x, y) = \frac{H(x)^2}{2\eta}\frac{\partial \mathcal{P}}{\partial x}\frac{y}{H(x)}\left(\frac{y}{H(x)} - 1\right) + V\left(1 - \frac{y}{H(x)}\right). \qquad (5.3)$$

Similarly,

$$q = -\frac{H(x)^3}{12\eta}\frac{\partial \mathcal{P}}{\partial x} + \frac{VH(x)}{2}. \qquad (5.4)$$

Because we are dealing with incompressible liquids, q is a constant for all x. Unlike flow between parallel planes, however, where we showed that the pressure gradient $\partial \mathcal{P}/\partial x$ is a constant, we see from Equation 5.4 that $\partial \mathcal{P}/\partial x$ is itself a function of x; indeed, we may consider Equation 5.4 to be an equation that *defines* the spatial distribution of pressure, which we know is not linear, and write

$$\frac{\partial \mathcal{P}}{\partial x} = 12\eta\left[\frac{V}{2H(x)^2} - \frac{q}{H(x)^3}\right]. \qquad (5.5)$$

Equation 5.3 can then be rewritten, after some algebraic manipulation, as

$$v_x(x, y) = V\left(1 - \frac{y}{H(x)}\right)\left[1 - 3\left(1 - \frac{2q}{VH(x)}\right)\frac{y}{H(x)}\right]. \qquad (5.6)$$

v_x will take on negative values in a region adjacent to the upper surface whenever $H(x) > 3q/V$.

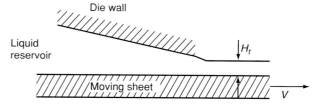

Figure 5.2. Schematic of a sheet-coating die.

Equation 5.5 can be solved for the pressure as a function of position by integrating both sides with respect to x to obtain[*]

$$\mathcal{P}(x) = \mathcal{P}_o + 12\eta \int_0^x \left[\frac{V}{2H(x)^2} - \frac{q}{H(x)^3} \right] dx. \tag{5.7}$$

$\mathcal{P}_o$ is a constant of integration that equals the pressure at $x = 0$. Finally, we obtain a useful expression relating the overall pressure change $\mathcal{P}_o - \mathcal{P}(L)$, the flow rate q, and the relative velocity V by setting $x = L$ in Equation 5.7 and rearranging:

$$q = \frac{\mathcal{P}_o - \mathcal{P}(L)}{12\eta \int_o^L H(x)^{-3} dx} + \frac{V \int_0^L H(x)^{-2} dx}{2 \int_0^L H(x)^{-3} dx}. \tag{5.8}$$

5.3 Sheet Coating

A process for polymer coating of a sheet is shown schematically in Figure 5.2. The schematic is the same for wire coating, except that we then understand the figure to represent a section taken through the axisymmetric cylindrical geometry. The analysis for sheet coating applies directly to wire coating if the maximum spacing between the wire and the die wall is small compared to the radius of the wire.

The coating thickness H_t, multiplied by the speed of the sheet, must equal the flow rate per unit width, q:

$$q = V H_t. \tag{5.9}$$

---

[*] The integration is with respect to x at fixed y because we are integrating a partial derivative. Thus, we might expect to find an additional term in Equation 5.7 that depends only on y. We can show that such a term will be negligible. Equation 3.4 will still be correct to within an error of order dH/dx, so we may expect any y dependence of $\mathcal{P}$ to be equal to τ_{yy}. But $\tau_{yy} = 2\eta\partial v_y/\partial y = $ (from the continuity equation) $-2\eta\partial v_x/\partial x$. From Equation 5.6 we may then write, after some simplification,

$$\tau_{yy} = -2\eta \frac{\partial v_x}{\partial x} = \frac{2\eta V}{H(x)} \left(1 - \frac{3q}{V H(x)} \right) \frac{dH}{dx}.$$

Thus, τ_{yy} is independent of y in this approximation and is in any event of order dH/dx, which we assume to be small.

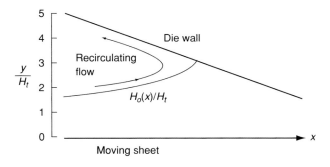

Figure 5.3. Recirculating flow in a sheet-coating die.

The pressure at the die exit, $\mathcal{P}(L)$, is atmospheric and can be taken as equal to zero. Equation 5.8 can then be written

$$H_t = \frac{\mathcal{P}_o}{12\eta V \int\limits_o^L H(x)^{-3}dx} + \frac{1}{2}\frac{\int\limits_0^L H(x)^{-2}dx}{\int\limits_0^L H(x)^{-3}dx}. \tag{5.10}$$

The coating thickness depends only on the die geometry if there is no net pressure drop through the die (i.e., the reservoir is at atmospheric pressure and $\mathcal{P}_o = 0$). We gain a degree of freedom, however, by pressurizing the reservoir, perhaps by feeding the polymer with an extruder. The reservoir pressure required for a given film thickness can be obtained by rewriting Equation 5.10 as

$$\mathcal{P}_o = 12\eta V \int\limits_o^L \frac{1}{H(x)^2}\left[\frac{H_t}{H(x)} - \frac{1}{2}\right]dx. \tag{5.11}$$

It follows from Equation 5.11 that the coating thickness of a Newtonian liquid can never be less than one half the exit gap of a converging die, or else a negative pressure drop would be required. [$H_t/H(L) < 1/2$ is in fact sufficient but not necessary for a negative reservoir pressure; the actual condition is more stringent and depends on the die shape $H(x)$.]

By replacing q with V/H_t in Equation 5.6, we obtain

$$v_x(x, y) = V\left(1 - \frac{y}{H(x)}\right)\left[1 - 3\left(1 - \frac{2H_t}{H(x)}\right)\frac{y}{H(x)}\right]. \tag{5.12}$$

There will be a region of negative velocity whenever $H(x) > 3H_t$. The situation in the die is shown in Figure 5.3, where the wall is taken to be planar. There is a region of zero net flow (the forward flow is exactly compensated for by the reverse flow) in the region $H_o \leq y \leq H$, where H_o is defined by the equation

$$\int\limits_{H_o}^{H} v_x dy = 0. \tag{5.13}$$

Figure 5.4. Schematic of two-dimensional flow between closely spaced surfaces.

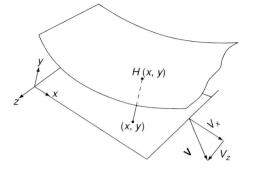

Substituting Equation 5.12 into the integral and rearranging the result leads to the explicit equation for H_o:

$$H_o(x) = \frac{H(x)H_t}{H(x) - 2H_t}. \tag{5.14}$$

The melt in the region $y < H_o(x)$ is swept from the die and forms the coating, while liquid in the region $H_o \leq y \leq H(x)$ simply recirculates. Good die design will avoid recirculation, since the recirculating liquid will have a long residence time at high temperature and degradation of the coating material could result.

This analysis can clearly be repeated for a power-law or other non-Newtonian viscosity function, as discussed in Section 3.4, but we would gain little additional insight here to justify the considerable algebraic complexity. In practice, more complex situations would be handled numerically, as discussed in Chapter 8.

5.4 Two-Dimensional Flows

Now suppose we have a thin gap geometry – a mold, for example – in which flow occurs in both the x and z directions. The geometry is shown schematically in Figure 5.4, where the x–z coordinate system is embedded in the lower surface and the upper surface is located at a distance $y = H(x,z)$. The lower surface moves relative to the upper with a velocity V; V is a vector with components V_x and V_z in the x and z directions, respectively. Note that we could always align the coordinates in the x–z plane so that either V_x or V_z is zero but, as we saw with the screw extruder, this might not be the most physically meaningful choice.

Consistent with the lubrication approximation, we assume that the flow is locally parallel to the surface at $y = 0$, and we neglect the component v_y of the velocity vector $\mathbf{v}$. The continuity equation is then

$$\frac{\partial v_x}{\partial x} + \frac{\partial v_z}{\partial z} = 0. \tag{5.15}$$

We will consider only the Newtonian fluid for simplicity. The components of the creeping flow equation then become

$$x: \quad \frac{\partial \mathcal{P}}{\partial x} = \eta \frac{\partial^2 v_x}{\partial y^2}, \tag{5.16a}$$

$$y: \quad \frac{\partial \mathcal{P}}{\partial y} = 0, \tag{5.16b}$$

$$z: \quad \frac{\partial \mathcal{P}}{\partial z} = \eta \frac{\partial^2 v_z}{\partial y^2}. \tag{5.16c}$$

We are assuming that the flow is fully developed at each value of x and z. $\mathcal{P}$ is independent of y, so we may integrate the x and z components to obtain

$$v_x(y) = V_x \left(1 - \frac{y}{H}\right) + \frac{yH}{2\eta} \left(\frac{\partial \mathcal{P}}{\partial x}\right) \left(\frac{y}{H} - 1\right), \tag{5.17a}$$

$$v_z(y) = V_z \left(1 - \frac{y}{H}\right) + \frac{yH}{2\eta} \left(\frac{\partial \mathcal{P}}{\partial z}\right) \left(\frac{y}{H} - 1\right). \tag{5.17b}$$

We have used the boundary conditions $v_x = v_z = 0$ at $y = H(x, z)$ and $v_x = V_x$, $v_z = V_z$ at $y = 0$. We integrate with respect to y from $y = 0$ to $y = H$ to obtain

$$q_x = \int_0^H v_x dy = \frac{V_x H}{2} + \frac{H^3}{12\eta} \left(-\frac{\partial \mathcal{P}}{\partial x}\right), \tag{5.18a}$$

$$q_z = \int_0^H v_z dy = \frac{V_z H}{2} + \frac{H^3}{12\eta} \left(-\frac{\partial \mathcal{P}}{\partial z}\right). \tag{5.18b}$$

$H = H(x, z)$ in all these equations.

Equations 5.17 and 5.18 are formally equivalent to those we used for the unwrapped model of the screw extruder, and they provide the starting point for considering cases with a channel depth that varies with x. It is often useful to employ a different formulation when H varies with both x and z, however, for in that case we cannot expect either q_x or q_z to remain constant. Now, we recall the Leibniz rule for differentiating an integral:

$$\frac{\partial q_x}{\partial x} = \frac{\partial}{\partial x} \int_0^{H(x,z)} v_x dy = \int_0^{H(x,z)} \frac{\partial v_x}{\partial x} dy + v_x(H) \frac{\partial H}{\partial x}. \tag{5.19}$$

The second term vanishes in our application because of the boundary condition $v_x = 0$ at $y = H$. There is a similar equation for $\partial q_z/\partial z$, and by summing the two equations we obtain

$$\frac{\partial q_x}{\partial x} + \frac{\partial q_z}{\partial z} = \int_0^H \left(\frac{\partial v_x}{\partial x} + \frac{\partial v_z}{\partial z}\right) dy = 0, \tag{5.20}$$

from which it follows that

$$\frac{\partial}{\partial x} \left(H^3 \frac{\partial \mathcal{P}}{\partial x}\right) + \frac{\partial}{\partial z} \left(H^3 \frac{\partial \mathcal{P}}{\partial z}\right) = 6\eta \left[\frac{\partial H}{\partial x} V_x + \frac{\partial H}{\partial z} V_z\right]. \tag{5.21}$$

Equation 5.21, which is the Reynolds lubrication equation, is a linear partial differential equation for $\mathcal{P}(x,z)$ of a type that is known as *elliptic* and whose properties are well understood. Elliptic equations require that the dependent variable ($\mathcal{P}$ in our case), its derivative, or a linear combination of both be specified everywhere on the boundary of the flow domain. It is likely that $\mathcal{P}$ will be specified at inflow and outflow boundaries. Walls through which there is no flow will be characterized by q_x and/or $q_z = 0$; a boundary at which at $q_x = 0$, for example, will have boundary condition $\partial \mathcal{P}/\partial x = 6\eta V_x/H^2$. The velocity field can be determined from Equations 5.17a–b once the function $\mathcal{P}(x,z)$ is known, after which we can calculate the stress, power, and so forth. We will not use Equation 5.21 in our applications, but it is often introduced as the starting point for a lubrication approximation analysis, and it is important to recognize that it is equivalent to the formulation that we have developed and employed here.

BIBLIOGRAPHICAL NOTES

Our starting point for the lubrication approximation was intuitive, where we assumed that flow in a nearly parallel channel would approximate flow in a channel with a slowly varying cross section to within terms of order α. It can be shown that the approximate is valid provided $\alpha \rho V H/\eta \ll 1$, where H is a characteristic gap spacing. This is a less restrictive condition than creeping flow, which requires $\rho V H/\eta \ll 1$; since $\alpha \ll 1$, the Reynolds number need not be small. The development of the ordering analysis can be found in many fluid mechanics books, including

Batchelor, G. K., *An Introduction to Fluid Dynamics*, Cambridge University Press, London, 1967, pp. 217ff.
Denn, M. M., *Process Fluid Mechanics*, Prentice Hall, Englewood Cliffs, NJ, 1980, ch. 13.
Sherman, F. S., *Viscous Flow*, McGraw-Hill, New York, 1990, pp. 229ff.

The sheet-coating example in this chapter is from Denn. See also the following texts on polymer processing, where the lubrication approximation is applied to a variety of processing flows in subsequent chapters after it is introduced:

Middleman, S., *Fundamentals of Polymer Processing*, McGraw-Hill, New York, 1977, pp. 172ff.
Pearson, J. R. A., *Mechanics of Polymer Processing*, Elsevier Applied Science, London, 1985, ch. 8.
Tadmor, Z., and C. G. Gogos, *Principles of Polymer Processing*, 2nd ed., Wiley InterScience, New York, 2007, pp. 64ff.

Middleman introduces the approximation with an assumption of creeping flow. Pearson has a rather formal development following an intuitive introduction.

6 Quasi-Steady Analysis of Mold Filling

6.1 Introduction

The creeping flow approximation to the momentum equation, which we obtain in a formal way by setting $\rho = 0$ on the left side of the equations in Table 2.2 or 2.4, has the interesting property that time never appears explicitly. Thus, creeping flow solutions to time-dependent problems are *quasi-steady* in the sense that they correspond to the steady-state solution for the given geometry at each time. This property can be exploited to obtain analytical solutions to simple transient problems in mold filling, and the same concepts are utilized for numerical solutions to more complex problems. We illustrate the use of the quasi-steady character of the creeping flow equations with two model mold filling problems, one in injection molding and one in compression molding.

6.2 Center-Gated Disk Mold

6.2.1 Isothermal Newtonian Liquid

A mold to form a thin circular disk is shown in Figure 6.1. Molten polymer is fed through a small circular hole at the center of the mold (the *gate*) and then flows out radially to fill the mold cavity. We assume that the mold is vented, allowing air to escape as the polymer fills the cavity, so the pressure at the polymer/air interface is always close to atmospheric. The disk has a thickness H and a radius R_D. The radius of the circular gate is R_G. The pressure at the gate is $\mathcal{P}_o$, and the polymer enters with a volumetric flow rate Q; $\mathcal{P}_o$ and Q may vary with time.

We assume isothermal flow and a Newtonian fluid. (Isothermality is the more serious of the two assumptions. Even if the physical properties of the melt are independent of temperature, we must deal with the possibility of solidification of the melt near the cold mold faces during filling.) For radial flow we have a single velocity component in cylindrical coordinates, v_r, and the continuity equation from Table 2.1 is

$$\frac{1}{r}\frac{\partial}{\partial r}r v_r = 0. \tag{6.1}$$

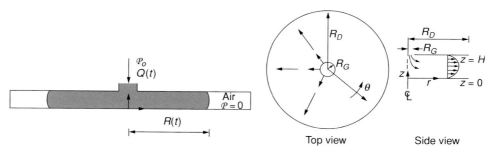

Figure 6.1. Schematic of filling a center-gated mold.

rv_r must be independent of r since the rate of change of rv_r with respect to r is zero; that is,

$$rv_r = f(z). \tag{6.2}$$

The components of the creeping flow equations are obtained from Table 2.4 by setting $\rho = 0$; with the velocity of the form $v_r = f(z)/r$, $v_\theta = v_z = 0$, we obtain

$$r \text{ component:} \quad \frac{\partial \mathcal{P}}{\partial r} = \eta \frac{\partial^2 v_r}{\partial z^2} = \frac{\eta}{r} \frac{d^2 f}{dz^2}, \tag{6.3a}$$

$$z \text{ component:} \quad \frac{\partial \mathcal{P}}{\partial z} = 0. \tag{6.3b}$$

Thus, $\mathcal{P}$ is a function only of r (and perhaps of time, which does not appear explicitly). Equation 6.3a can be rewritten as

$$\eta \frac{d^2 f}{dz^2} = r \frac{d\mathcal{P}}{dr} = \frac{d\mathcal{P}}{d \ln r} = C_1. \tag{6.4}$$

Since z and r are independent variables, the function of z on the left of Equation 6.4 can equal the function of r on the right for all r and z only if both are constant.

The function $f(z)$ must be a quadratic, since its second derivative is a constant. After two integrations we obtain

$$f(z) = \frac{C_1}{2\eta} z^2 + C_2 z + C_3. \tag{6.5}$$

v_r will vanish at the upper and lower mold faces, $z = 0$ and $z = H$, so we must have $f(0) = f(H) = 0$. It readily follows that $C_3 = 0$ and $C_2 = -C_1 H/2\eta$, so

$$f(z) = \frac{C_1}{2\eta}(z^2 - zH). \tag{6.6}$$

The differential flow rate through a cylindrical section of height dz and radius r is $2\pi r v_r dz$ ($2\pi r dz$ is the differential area); the total flow rate at any radius r is therefore

$$Q = \int_0^H 2\pi r v_r dz = \int_0^H 2\pi f(z)dz = -\frac{C_1 \pi H^3}{6\eta}, \tag{6.7}$$

which establishes the value of C_1 as $-6\pi Q/\pi H^3$.

The pressure is obtained from Equation 6.4 as

$$P(r) = P_o + C_1 \ln \frac{r}{R_G} = P_o - \frac{6\eta Q}{\pi H^3} \ln \frac{r}{R_G}. \tag{6.8}$$

Suppose the mold is filled to a radius $R(t)$ at time t; the pressure $P(R)$ is zero because the mold is vented, so we have a relation between the instantaneous fill pressure, the instantaneous flow rate, and the amount of material in the mold:

$$P_o(t) = \frac{6\eta Q}{\pi H^3} \ln \frac{R(t)}{R_G}. \tag{6.9}$$

It is a consequence of the quasi-steady description that time enters only implicitly.

Now, suppose we are filling the mold at constant pressure P_o and wish to know the time required to fill the cavity completely ($R = R_D$). The filled volume at any time is $\pi R^2 H$. The flow rate equals the rate of change of the volume, so

$$Q = \frac{d(\pi R^2 H)}{dt} = 2\pi R \frac{dR}{dt} H. \tag{6.10}$$

Equation 6.10 can then be written

$$R \ln \left(\frac{R}{R_G} \right) \frac{dR}{dt} = \frac{P_o H^2}{12\eta} \tag{6.11a}$$

or

$$\int_{R_G}^{R_D} R \ln \left(\frac{R}{R_G} \right) dR = \int_0^{t_D} \frac{P_o H^2}{12\eta} dt. \tag{6.11b}$$

The integral on the left is a standard form that can be integrated by parts or found in a table of integrals, and we obtain the fill time as

$$t_D = \frac{3\eta}{P_o H^2} \left[2R_D^2 \ln \left(\frac{R_D}{R_G} \right) - (R_D^2 - R_G^2) \right]. \tag{6.12}$$

The analysis is valid only for $R_D \gg R_G$, so we obtain, finally,

$$t_D \approx \frac{3\eta R_D^2}{P_o H^2} \left[2 \ln \left(\frac{R_D}{R_G} \right) - 1 \right]. \tag{6.13}$$

$\ln(R_D/R_G)$ is a weak function of its argument; it varies only from 2.3 for $R_D/R_G = 10$ to 3.9 for $R_D/R_G = 50$, so the estimated fill time is relatively insensitive to the radius of the gate.

It is clear that we have neglected all details of the flow near the free surface, including any curvature. The flow near the three-phase *moving contact line*, where the melt, the mold wall, and the air all meet, is very difficult to analyze with the no-slip boundary condition and is not completely understood.* The time to fill the mold

---

* As the mold fills, liquid must wet the surface of the mold. According to the no-slip boundary condition, once a fluid element is in contact with the solid surface it must remain in the same place, so fluid at the mold surface cannot be moving outward along the surface. Hence, the fluid covering the surface of the mold must come from the interior. It is possible to construct kinematics that permit fluid to "roll" in a way that the contact line moves outward while the no-slip boundary condition is satisfied, but this rolling motion leads to a stress that, when integrated over any region including

isothermally is insensitive to the details of this flow, but the flow near the contact line has a significant effect on the orientation of polymer chains and on the temperature profile under nonisothermal conditions.

6.2.2 Isothermal Power-Law Fluid

The extension to non-Newtonian fluids, for which the viscosity is a function of $\frac{1}{2}II_D$ (Table 2.5), is not straightforward. With $v_r = f(z)/r$, $v_\theta = v_z = 0$, the invariant has the form

$$II_D = \frac{2}{r^2}\left(\frac{df}{dz}\right)^2 + \frac{f^2}{r^4}. \tag{6.14}$$

Since there are two different powers of r in the functional dependence of the viscosity, it will clearly be impossible to obtain an equation for $f(z)$ in the manner used for the Newtonian liquid.

There is an approximate approach, in the spirit of the lubrication approximation, that can be used, and we will illustrate it for the power-law fluid. We expect that over most of the mold volume the stretch rate $\partial v_r/\partial r$ will be substantially less than the shear rate $\partial v_r/\partial z$. In that case, the second term on the right of Equation 6.14 becomes negligible relative to the first, and some small terms vanish from the creeping flow equations. The r component of the momentum equation in the creeping flow approximation then becomes

$$\frac{\partial \mathcal{P}}{\partial r} = \frac{\partial \tau_{rz}}{\partial z} = \frac{\partial}{\partial z} K \left|\frac{f'^2}{r^2}\right|^{\frac{n-1}{2}} \frac{f'}{r} \tag{6.15a}$$

We make use of the symmetry about the center plane to restrict the analysis to $0 \le z \le H/2$, where $f'(z) \ge 0$. The boundary conditions on f are

$$f(0) = f'\left(\frac{H}{2}\right) = 0. \tag{6.15b}$$

Equation 6.15a can be written

$$r^n \frac{\partial \mathcal{P}}{\partial r} = K \frac{d}{dz}(f')^n = -C_1^n, \tag{6.16}$$

where $C_1 > 0$ is a constant. This equation is integrated once with respect to z and rearranged to obtain

$$f' = \frac{C_1}{K^n}\left(\frac{H}{2} - z\right)^{\frac{1}{n}}, \quad 0 \le z \le \frac{H}{2}, \tag{6.17}$$

where we have employed the condition $f'(H/2) = 0$. A second integration gives

$$f(z) = \frac{n}{n+1}\frac{C_1}{K^{1/n}}\left(\left(\frac{H}{2}\right)^{\frac{n+1}{n}} - \left(\frac{H}{2} - z\right)^{\frac{n+1}{n}}\right), \quad 0 \le z \le \frac{\pi}{2}. \tag{6.18}$$

the interface, produces an infinite force. (Infinite stresses at a point are permissible, but they must integrate to finite forces.) The usual practice in numerical schemes is to relax the no-slip condition in a very small region near the contact line. Computed results in the remainder of the mold do not depend on the details of the way in which the no-slip condition is modified.

Finally, we obtain the flow rate Q from

$$Q = \int_0^H 2\pi r v_r dz = 2 \int_0^{H/2} 2\pi r v_r dz = \frac{\pi n C_1 H^{\frac{2n+1}{n}}}{2^{1/n}(2n+1)}. \tag{6.19}$$

This gives the constant C_1 in terms of Q. The pressure profile is then determined from Equation 6.16, and the time to fill the mold for a given gate pressure is found as in the preceding section.

    The key question is "How good an approximation is it to neglect the terms involving $\partial v_r / \partial z$?" We can get an estimate of the error by evaluating the ratio of the two terms in Equation 6.14 as follows:

$$\Lambda = \frac{f^2}{r^4} \Big/ \frac{2f'^2}{r^2} = \frac{1}{2r^2} \left(\frac{n}{n+1}\right)^2 \frac{\left(\left(\frac{H}{2}\right)^{\frac{n+1}{n}} - \left(\frac{H}{2} - z\right)^{\frac{n+1}{n}}\right)^2}{\left(\frac{H}{2} - z\right)^{\frac{2}{n}}}. \tag{6.20a}$$

Consider any position $z = \lambda H/2,\ 0 \leq \lambda \leq 1$. Then

$$\Lambda = \frac{1}{8}\left(\frac{n+1}{n}\right)^2 \left(\frac{H}{r}\right)^2 \left[\frac{1 - (1-\lambda)^{\frac{n+1}{n}}}{(1-\lambda)^{\frac{1}{n}}}\right]^2. \tag{6.20b}$$

Λ is small and the approximation is good provided $H/r \ll 1$, where the actual magnitude depends on λ. The approximation must fail on the center plane, where the shear stress vanishes and $\Lambda \to \infty$, but this should have little effect on the overall result. We expect considerable error during the early stages of mold filling, when H/r is not small, but we should obtain a reasonably accurate estimate of the total fill time.

6.2.3 Coupled Flow

The analysis of slit flow in Chapter 4, where the consideration of variable physical properties and heat transfer was shown to introduce major qualitative changes, is even more relevant to injection molding, where the pressures and flow rates are high. The analog of Equation 6.9 for a pressure-dependent viscosity is

$$\mathcal{P}_o = \frac{1}{\beta} \ln\left(1 + \frac{6\beta\eta^o Q}{\pi H^3} \ln \frac{R}{R_G}\right). \tag{6.21}$$

The cross-sectional averaging to obtain the analysis leading to Equation 4.20 for developing flow with heat transfer is a bit more delicate because of the radial dependence of v_r, but the procedure is the same and no new phenomena are expected.

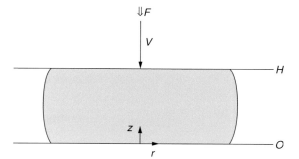

Figure 6.2. Schematic of compression molding.

6.3 Closing a Compression Mold

6.3.1 Isothermal Newtonian Fluid

A schematic of a simple compression molding process is shown in Figure 6.2. A polymer charge is contained between two mold faces, one of which moves toward the other. The polymer fills the space as the mold faces close. For simplicity we assume that the mold faces are flat circular disks with radius R_D, and we assume that the initial polymer charge is a cylinder with height H_o and radius R_o. We can therefore assume axisymmetry in the flow. The upper face moves toward the lower with a velocity V, and the force required to close the mold is F; both V and F can be functions of time.

The key to analyzing the flow in the mold is recognizing that layers of fluid are stretched radially as the mold is closed. We assume that the stretching is uniform; that is, if we mark a fluid layer of constant thickness, the layer will move toward the stationary plate uniformly and stretch radially such that it will still be of constant (smaller) thickness. This assumption is equivalent to the kinematic assumption that v_z is a function only of z; we add the assumption of no circumferential flow and take as a starting point

$$v_z = \phi(z), \quad v_\theta = 0. \tag{6.22}$$

The equation of continuity in cylindrical coordinates can then be written

$$\frac{1}{r}\frac{\partial}{\partial r}r v_r = -\frac{d\phi(z)}{dz}. \tag{6.23}$$

This is integrated to

$$r v_r = -\frac{r^2}{2}\frac{d\phi}{dz} + C. \tag{6.24a}$$

C can be a function of z, but it must vanish if we require that v_r remain finite as $r \to 0$. Thus, the radial velocity is of the form

$$v_r = -\frac{r}{2}\frac{d\phi}{dz}. \tag{6.24b}$$

We now assume isothermal creeping flow of a Newtonian liquid. The r and z components of the creeping flow equations, with the assumed form for the velocity, are

$$r \text{ component:} \quad \frac{\partial \mathcal{P}}{\partial r} = -\frac{1}{2}\eta r \frac{d^3\phi}{dz^3}, \tag{6.25a}$$

$$z \text{ component:} \quad \frac{\partial \mathcal{P}}{\partial z} = \eta \frac{d^2\phi}{dz^2}. \tag{6.25b}$$

We can eliminate the pressure by differentiating the first equation with respect to z and the second with respect to r, then equating $\partial^2 \mathcal{P}/\partial z \partial r$ to $\partial^2 \mathcal{P}/\partial r \partial z$; the result will be to show that $d^4\phi/dz^4 = 0$. We gain more insight into the structure of the solution, however, by first examining the equations. The right-hand side of Equation 6.25b is a function of z, so $\partial \mathcal{P}/\partial z$ depends only on z. It thus follows that $\mathcal{P}$ must be separable into the sum of a function of z and a function of r. In that case, $\partial \mathcal{P}/\partial r$ depends only on r, so it follows from Equation 6.25a that $d^3\phi/dz^3 = (2r/\eta)\partial \mathcal{P}/\partial r$ must be a constant, since that is the only way in which a function of z can equal a function of r. $\phi(z)$ must therefore be a cubic, with four coefficients to be determined.

The boundary conditions at the lower and upper plates, respectively, are

$$z = 0: \quad v_z = \phi(0) = 0, \quad v_r = -\frac{r}{2}\frac{d\phi(0)}{dz} = 0, \tag{6.26a}$$

$$z = H: \quad v_z = \phi(H) = -V, \quad v_r = -\frac{r}{2}\frac{d\phi(H)}{dz} = 0, \tag{6.26b}$$

or

$$z = 0: \quad \phi(0) = \frac{d\phi(0)}{dz} = 0, \tag{6.27a}$$

$$z = H: \quad \phi(H) = -V, \quad \frac{d\phi(H)}{dz} = 0. \tag{6.27b}$$

The velocity is then

$$v_z = \phi(z) = -3V\left(\frac{z}{H}\right)^2\left(1 - \frac{2}{3}\frac{z}{H}\right), \tag{6.28a}$$

$$v_r = -\frac{r}{2}\frac{d\phi}{dz} = \frac{3rzV}{H^2}\left(1 - \frac{z}{H}\right). \tag{6.28b}$$

Note that $H = H(t)$ and, in general, $V = V(t)$. We will subsequently make use of the fact that $dH/dt = -V$.

We will need a relationship between the flow field and the force required to close the mold. The force is obtained by integrating the stress σ_{zz} over the wetted surface at $z = H$. Since $\sigma_{zz} = -\mathcal{P} + \tau_{zz}$, we must compute the pressure. (We will not distinguish between $\mathcal{P}$ and p, since the maximum gravitational term $\rho g H$ will be negligible.) From Equation 6.25a it is obvious that $\mathcal{P}$ is quadratic in r ($\partial \mathcal{P}/\partial r$ is proportional to r), while from Equation 6.25b it is obvious that $\mathcal{P}$ is quadratic in z;

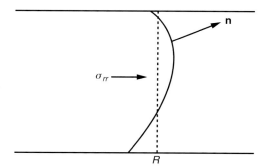

Figure 6.3. Liquid–air interface in a compression mold.

the coefficients are easily obtained by substituting the function $\phi(z)$ into Equations 6.25a–b, leading to

$$\mathcal{P} = \mathcal{P}_o + \frac{3\eta V}{H}\left[2\frac{z}{H}\left(\frac{z}{H} - 1\right) - \frac{r^2}{H^2}\right]. \tag{6.29}$$

$\mathcal{P}_o$ is a constant of integration.

Determination of $\mathcal{P}_0$ is rather delicate and illustrates a problem that is seen frequently in systems with polymer/air interfaces. We have not yet specified any boundary conditions in the radial direction, despite the finite size of the system. This is because the kinematics we have assumed ignore the finite size; hence, they ignore any changes in the flow in the neighborhood of the free surface. The free surface is shown in Figure 6.3, where the shape of the surface has been exaggerated to make a point. The location R of the edge of the melt needs to be defined; we can think of it as the average over z, although some weighting of the average might be appropriate. The stress normal to the interface at any point will differ from atmospheric pressure only by an amount equal to the surface tension multiplied by the sum of the reciprocals of the two radii of curvature, and the surface tension contribution will always be negligible relative to the viscous stresses for polymer melts, so we may take the stress normal to the surface to be equal to zero. This is not a helpful observation, however, because we have no knowledge of the details of the flow near the interface; hence, there is no way to calculate either the normal direction **n** or the components of the stress.

If we could take the interface to be a cylinder (i.e., a liquid–solid contact angle of $90°$), then our free surface condition would be $\sigma_{rr} = -\mathcal{P} + \tau_{rr} = 0$. This approximation is also inadequate in the present context, however, since if we take the assumed kinematics to be valid right up to $r = R$ we will find that σ_{rr} is a function of z and cannot vanish everywhere. The best resolution at this level of approximation is to take the interface to be a cylinder, to take the kinematics in Equations 6.28a–b to be valid at $r = R$, and to set the *average* value of σ_{rr} at $r = R$ equal to zero. Thus, we determine $\mathcal{P}_o$ from the equation

$$\frac{1}{H}\int_0^H \sigma_{rr}dz = \frac{1}{H}\int_0^H (-\mathcal{P} + \tau_{rr})dz = 0 \quad \text{at } r = R. \tag{6.30}$$

$\mathcal{P}$ is obtained from Equation 6.29 and $\tau_{rr} = 2\eta \partial v_r/\partial r$ from Equation 6.28b; the result is

$$\mathcal{P}_o = \frac{2\eta V}{H} + \frac{3\eta V}{H}\left(\frac{R}{H}\right)^2. \tag{6.31}$$

To calculate the force we need to integrate $\sigma_{zz} = -\mathcal{P} + \tau_{zz}$ over the moving surface at $z = H$. It readily follows that $\tau_{zz} = 2\eta \partial v_z/\partial z = 0$ at $z = H$, and the term involving z in Equation 6.29 also vanishes. We thus calculate the force as

$$F = 2\pi \int_0^R \mathcal{P}(r, H) r \, dr = 2\pi \int_0^R \left[\frac{3\eta V}{H^3}(R^2 - r^2) + \frac{2\eta V}{H}\right] r \, dr$$

$$= \frac{3\pi \eta V R^4}{2H^3}\left[1 + \frac{4}{3}\left(\frac{H}{R}\right)^2\right]. \tag{6.32}$$

We are generally concerned with $H \ll R$, so we may neglect the second term* and write

$$F \approx \frac{3\pi \eta V R^4}{2H^3}. \tag{6.33}$$

At this point we need to consider the mode of operation of the molding machine: controlled speed, controlled force, or some combination of the two. If the speed of the mold surface is controlled, V and H are known functions of time and $R^2 = R_o^2 H_o/H$, so F is known for all time until the mold is closed. We will consider here the case in which the closing *force* is kept constant. Since $V = -dH/dt$ we can then rewrite Equation 6.33 as

$$-\frac{dH}{dt} = \frac{2FH^3}{3\pi \eta R^4} = \frac{2FH^5}{3\pi \eta R_o^4 H_o^2}, \tag{6.34}$$

where we have used $R^2 H = R_o^2 H_o$. This separable equation can be integrated immediately to obtain

$$\frac{1}{4}\left(H^{-4} - H_o^{-4}\right) = \frac{2Ft}{3\pi \eta R_o^4 H_o^2}. \tag{6.35}$$

We will generally have $H_o^{-4} \ll H^{-4}$, so Equation 6.35 can be rearranged to give an estimate of the time t_D to fill the mold to radius R_D as

$$t_D \approx \frac{3\pi \eta R_D^8}{8F R_o^4 H_o^2}. \tag{6.36}$$

The very strong dependence on the radius of the mold is a reflection of the rapid decrease in velocity at fixed force as H gets small (cf. Equation 6.34).

---

* The equivalent of the second term can be very important if the viscosity varies across the gap. Such a situation would arise with a bicomponent charge, for example, in which there is a central layer of very high viscosity sandwiched between outer layers of low viscosity. The traditional treatment of the edge boundary condition in the fluid mechanics and lubrication literature differs from the approach used here and does not lead to the second term in Equation 6.32.

6.3.2 Isothermal Power-Law Fluid

The treatment of non-Newtonian fluids is analogous to that in Section 6.2.2 for the center-gated disk mold. It is assumed that the function II_D is dominated by the term $\partial v_r/\partial z = -\frac{1}{2}d\phi(z)/dz$ and that terms involving $\partial v_r/\partial r$ can be neglected in general; in that case an ordinary differential equation can be obtained for $\phi(z)$, and the solution for a power-law fluid follows the procedure in the preceding section. There is nothing qualitatively different from the result for a Newtonian fluid, but of course the velocity, pressure, force, and fill time now depend on the power-law exponent n.

6.3.3 Nonisothermal Flow

The analysis of nonisothermal flow in a compression mold is complex for two reasons. The first is that the quasi-steady approximation that is inherent in the creeping flow equations does not carry over to the energy equation, so the full transient equation for the temperature must be solved in two spatial dimensions, where the spatial regime is changing with time. The second is that as in the problem of filling an injection mold, the "fountain flow" near the free surface is the source of the fluid that coats the mold face as the mold closes and the radius of the charge increases. Thus, fluid from the center of the flow is carried to the wall, and this fluid will have a different temperature from the fluid in the wall region. The deviation from the assumed velocity profile near the front is therefore an important factor in determining the temperature distribution.

Some useful insight can be obtained by considering the special case in which the viscosity is a function of z but independent of r; this situation might arise, for example, during the first moments after a cold charge is put into a mold with hot faces. If the viscosity of the fluid near the mold face is much lower than the viscosity of the fluid along the center plane of the mold, we might expect the near-wall fluid to flow more quickly and encapsulate the more viscous center fluid. It is straightforward to show, however, that with the assumed kinematics the radial velocity must increase monotonically from the wall to the center plane, thus precluding encapsulation. Numerical solutions of the type described in Chapter 8 show that encapsulation is possible under some conditions, and a lubrication-like approximate solution is possible. The structure of the solution is determined by the dimensionless group $S = \eta_{max} H^2/\eta_{min} R^2$, where η_{max} and η_{min} are the maximum and minimum viscosities in the system, respectively. Encapsulation is possible when $S \gg 1$, and the analytical solution for this condition agrees very well with the numerical simulation.

6.4 Concluding Remarks

The main point of this chapter has been to demonstrate the way in which the quasi-steady creeping flow equations can be used to solve transient mold filling problems in polymer processing. The same logic can be used for numerical solutions, wherein the flow field is determined at each time from the creeping flow equations for the given geometry. The new domain occupied by the polymer at the next time step is then determined from the velocity at the moving surface.

BIBLIOGRAPHICAL NOTES

The compression molding analysis follows that in Chapter 12 of

Denn, M. M., *Process Fluid Mechanics*, Prentice Hall, Englewood Cliffs, NJ, 1980.

The general solution for squeeze flow of a Newtonian fluid between finite plates of arbitrary shape is given in

Denn, M. M., and G. Marrucci, *J. Non-Newtonian Fluid Mech.*, **87**, 175 (1999).

The issues of the proper radial boundary condition for the pressure and the effect of a transverse viscosity gradient are treated in

Lee, S. J., M. M. Denn, M. J. Crochet, and A. B. Metzner, *J. Non-Newtonian Fluid Mech.*, **10**, 3 (1982).

The solution for a power-law fluid may be found in

Tadmor, Z., and C. G. Gogos, *Principles of Polymer Processing,* 2nd ed., Wiley InterScience, New York, 2007, pp. 291ff.

The approach used by Tadmor and Gogos, which is the conventional one in the literature, leads to a radial stress that is independent of z and will give the incorrect force for a fluid with a transverse viscosity gradient (which would be expected for a power-law fluid at high deformation rates).

A squeeze flow geometry is sometimes used for rheological measurement; see, for example,

Macosko, C. W., *Rheology: Principles, Measurements, and Applications*, VCH, New York, 1994, Sec. 6.4.4.

7 Fiber Spinning

7.1 Introduction

The processes we have considered thus far – extrusion, wire coating, and injection and compression molding – are dominated by *shear between* confined surfaces. By contrast, in fiber and film formation the melt is *stretched without* confining surfaces. It is still possible to gain considerable insight from very elementary flow and heat transfer models, but we must first parallel Section 2.2 and develop some basic concepts of extensional flow. The remainder of the chapter is then devoted to an analysis of fiber formation by melt spinning.

Our analysis of fiber spinning in this chapter will be based on an inelastic rheological model of the stresses. This rheological description appears to be adequate for polyesters and nylons, which comprise the bulk of commercial spinning applications, and our spinning model is essentially the one used in industrial computer codes. This is a process in which melt viscoelasticity can sometimes play an important role, however, and we will revisit the process in Chapter 10.

7.2 Uniaxial Extensional Flow

Consider a cylindrical rod of a very viscous polymer melt, as shown in Figure 7.1, with radius R and length L. We impose a stress σ_{zz} in the axial direction in order to stretch the rod; hence, R and L are both functions of time, but $R^2 L$ is a constant for an incompressible melt. We assume that the rod draws down uniformly as it is stretched, so R is independent of z. The cylindrical coordinate system is embedded at one end of the rod, so we may consider the end at $z = 0$ to be fixed and the end at $z = L$ to be moving.

One's first reaction is to wonder how such an experiment can be carried out. If suffices here to say that clever experimental designs for highly viscous materials have been implemented and even commercialized, although the experiment is a difficult one to do well. In fact, the first reported measurements, by Trouton, were done one hundred years ago, together with an analysis paralleling the one given here.

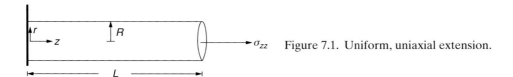

Figure 7.1. Uniform, uniaxial extension.

The primary kinematical assumption is that the axial component of velocity, v_z, is independent of r. Thus, layers of fluid at different distances from the axis do not move past one another, in which case no shear stresses are generated. We thus see the fundamental difference between this flow and the ones we have encountered previously. v_z will, however, depend on z; this is obvious since we have assumed that the end at $z = 0$ is fixed ($v_z = 0$ at $z = 0$), while the end at $z = L$ must have a finite velocity if the rod is to be extended.

Now, let us first apply the continuity equation. We assume axisymmetry ($\partial/\partial\theta = 0$) and no circumferential flow ($v_\theta = 0$). In cylindrical coordinates we thus have, from Table 2.1,

$$\frac{1}{r}\frac{\partial}{\partial r}(r v_r) + \frac{\partial v_z}{\partial z} = 0. \tag{7.1}$$

$\partial v_z/\partial z$ is independent of r, so we can rearrange Equation 7.1 and integrate with respect to r to obtain

$$r v_r = -\frac{r^2}{2}\frac{\partial v_z}{\partial z} + C_1. \tag{7.2}$$

v_z must remain finite as $r \to 0$, so the integration constant C_1 must equal zero and we have

$$v_r = -\frac{r}{2}\frac{\partial v_z}{\partial z}. \tag{7.3}$$

We assume that the filament is drawn down uniformly; that is, the radius must be independent of z. In that case, the radial velocity must be independent of z, since the filament will remain uniform only if the rate of thinning (the radial velocity) is the same everywhere. It therefore follows from Equation 7.3 that $\partial v_z/\partial z$ must be independent of z; we denote $\partial v_z/\partial z$ by $\dot{\gamma}_E$, the *rate of extension*. $\dot{\gamma}_E$ may be a function of time. We have assumed that $v_z = 0$ at $z = 0$, so it follows that the velocity field is of the form

$$v_z = \dot{\gamma}_E z, \qquad v_r = -\frac{r}{2}\dot{\gamma}_E. \tag{7.4a,b}$$

Note that this result is not dependent on the stress constitutive relation for the fluid.

Now, if there are no shear stresses, the r and z components of the creeping flow equations (Table 2.2 with $\rho = 0$) become

$$r: \quad 0 = -\frac{\partial \mathcal{P}}{\partial r} + \frac{1}{r}\frac{\partial}{\partial r}(r\tau_{rr}) - \frac{\tau_{\theta\theta}}{r} = -\frac{\partial \mathcal{P}}{\partial r} + \frac{\partial \tau_{rr}}{\partial r} + \frac{\tau_{rr} - \tau_{\theta\theta}}{r}, \tag{7.5a}$$

$$z: \quad 0 = -\frac{\partial \mathcal{P}}{\partial z} + \frac{\partial \tau_{zz}}{\partial z} = \frac{\partial}{\partial z}(-\mathcal{P} + \tau_{zz}). \tag{7.5b}$$

From Equation 7.5b we see that $-\mathcal{P} + \tau_{zz}$ is independent of z, although we know it will depend on time. Equation 7.5a can be written

$$\frac{\partial}{\partial r}(-\mathcal{P} + \tau_{rr}) = -\frac{\tau_{rr} - \tau_{\theta\theta}}{r}, \tag{7.6a}$$

or

$$-\mathcal{P} + \tau_{rr} = -\int_0^r \frac{\tau_{rr} - \tau_{\theta\theta}}{r} dr + C_2. \tag{7.6b}$$

The total radial normal stress inside the polymer cylinder (Equation 2.7), $\sigma_{rr} = -p + \tau_{rr}$, must be exactly balanced at $r = R$ by the pressure of the atmosphere (which we can take to be zero) plus the pressure change across the curved interface caused by the surface tension. It is almost always possible to neglect the surface tension contribution for polymer melts (but not for solutions!), and the radial gravitational contribution over the thin cylindrical cross section will be negligible, regardless of orientation, so we may ignore the distinction between $\mathcal{P}$ and p in Equation 7.5b. Thus, at $r = R$ we may take $-\mathcal{P} + \tau_{rr} = 0$ and write

$$C_2 = \int_0^R \frac{\tau_{rr} - \tau_{\theta\theta}}{r} dr \tag{7.6c}$$

or

$$\mathcal{P} = \tau_{rr} - \int_r^R \frac{\tau_{rr} - \tau_{\theta\theta}}{r} dr. \tag{7.7}$$

The total axial stress, σ_{zz}, is thus

$$\sigma_{zz} = -p + \tau_{zz} = -\mathcal{P} + \tau_{zz} + \rho g h = \tau_{zz} - \tau_{rr} + \int_r^R \frac{\tau_{rr} - \tau_{\theta\theta}}{r} dr + \rho g h. \tag{7.8}$$

The gravitational term is usually negligible for very viscous liquids, and we shall ignore it henceforth and write

$$\sigma_{zz} = \tau_{zz} - \tau_{rr} + \int_r^R \frac{\tau_{rr} - \tau_{\theta\theta}}{r} dr. \tag{7.9}$$

Equation 7.9 expresses the total axial stress in terms of the extra stress, so the stress constitutive equation provides the link to the kinematics.

We now assume that the melt is inelastic, although perhaps with a deformation-dependent viscosity. From Table 2.3, with v_z independent of r and v_r given by Equation 7.3, we then obtain

$$\tau_{rr} = 2\eta \frac{\partial v_r}{\partial r} = -\eta \frac{\partial v_z}{\partial z}, \tag{7.10a}$$

$$\tau_{\theta\theta} = 2\eta \frac{v_r}{r} = -\eta \frac{\partial v_z}{\partial z} = \tau_{rr}, \tag{7.10b}$$

$$\tau_{zz} = 2\eta \frac{\partial v_z}{\partial z}. \tag{7.10c}$$

The integral in Equation 7.9 vanishes because of the equality of τ_{rr} and $\tau_{\theta\theta}$. From Equations 7.10a and 7.10b we therefore obtain

$$\sigma_{zz} = \tau_{zz} - \tau_{rr} = 3\eta \frac{\partial v_z}{\partial z} = 3\eta \dot{\gamma}_E. \qquad (7.11)$$

Equation 7.11 is sometimes written

$$\sigma_{zz} = \eta_E \dot{\gamma}_E, \qquad (7.12)$$

where η_E is known as the *extensional viscosity* (and sometimes the *Trouton viscosity*). The ratio η_E/η is known as the *Trouton ratio*, and it is equal to 3 for inelastic liquids. If the viscosity is deformation rate dependent, the comparison must be made at the same value of II_D for consistency. The function $\sqrt{\frac{1}{2}II_D}$ in Table 2.6 is simply $\dot{\gamma}$ for a shear flow with $v_x = v_x(y)$ and $v_y = v_z = 0$, while for an extensional flow described by Equation 7.11 we obtain $\sqrt{\frac{1}{2}II_D} = \sqrt{3}\dot{\gamma}_E$. Thus, any comparison of extensional and shear viscosities should be made when $\dot{\gamma} = \sqrt{3}\dot{\gamma}_E$, not at equal values of $\dot{\gamma}$ and $\dot{\gamma}_E$.

Many highly elastic polymer melts, including polypropylene, which is spun into fibers and hence is directly relevant to the discussion here, show considerably more complex behavior in extension than that given by Equation 7.11. Equation 7.11 does seem to be adequate for many polymers that are spun commercially, however, including poly(ethylene terephthalate); hence, we will employ it in this chapter. (We use the qualifier "seem" because these polymers tend to be insufficiently viscous to carry out the extensional experiment as analyzed here, and the extensional behavior must be inferred from other measurements.) We therefore proceed with an analysis of spinning that, at the point where we require a constitutive equation, presumes the applicability of Equation 7.11.

7.3 Melt Spinning Equations

7.3.1 Formulation

The process for spinning a single fiber is shown schematically in Figure 7.2. Melt is extruded continuously through a small hole (the *spinneret*) into an ambient environment that is below the solidification temperature. Solidification occurs somewhere between the point of extrusion and a takeup device; takeup is at a much higher speed than the extrusion velocity, so the filament is drawn down in diameter as it passes through the melt zone. (We assume there is no permanent deformation of the solidified filament so that all of the area reduction occurs in the melt zone, which is not strictly true.) The commercial process usually has many spinnerets spaced across a single *spinneret plate*, all fed from the same *spinning head*, and the multiple filaments are gathered together into a *yarn* prior to takeup.

It is obvious that there is a strong interaction between the fluid dynamics and heat transfer in the melt zone. Our approach parallels the heat transfer analysis in Section 3.2.2 and is very similar in philosophy to the thin gap approximation

Temperature [°C]

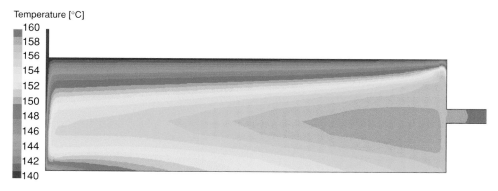

Figure 8.11. Computed temperature contours for a planar extruder.

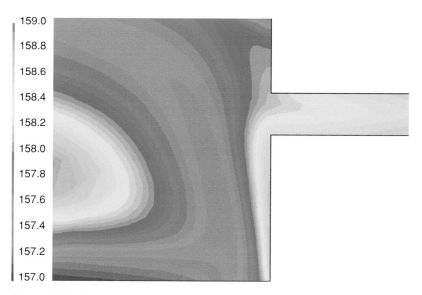

Figure 8.12. Computed temperature contours in the region near the die.

Figure 8.14. Computed temperature profile with uniform flow from the left.

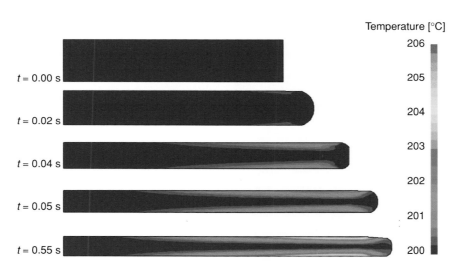

Figure 8.18. Shape and temperature evolution in time for adiabatic plates. Reprinted from Debbaut, *J. Non-Newtonian Fluid Mech.*, **98**, 15 (2001). Copyright Elsevier.

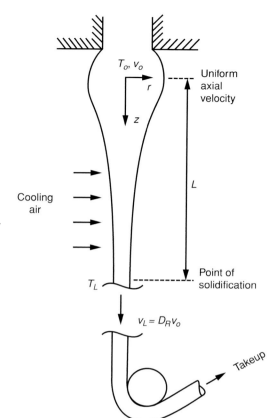

Figure 7.2. Schematic of melt spinning.

developed in Chapter 5, except that the flow field here is primarily extensional, not shear. The resulting model, known as the *thin filament equations*, can be derived in a number of ways, and we present a rather simpler but less informative derivation than the one used here in Appendix 7B.

We assume axisymmetry and a steady state. Commercial fiber spinning takes place at speeds on the order of 4,000 m/min (240 km/hr) and greater, so in this case inertia is important, as is aerodynamic drag; this is perhaps the only polymer melt process where inertia must be considered. We therefore use the full Cauchy momentum equations, and we write the basic equations in cylindrical coordinates (Tables 2.1, 2.2, and 2.5) as follows:

continuity:
$$\frac{1}{r}\frac{\partial}{\partial r}(r v_r) + \frac{\partial v_z}{\partial z} = 0, \tag{7.13}$$

r momentum:
$$\rho\left(v_r\frac{\partial v_r}{\partial r} + v_z\frac{\partial v_r}{\partial z}\right) = -\frac{\partial p}{\partial r} + \frac{1}{r}\frac{\partial}{\partial r}(r \tau_{rr}) - \frac{\tau_{\theta\theta}}{r} + \frac{\partial \tau_{rz}}{\partial z}, \tag{7.14a}$$

z momentum:
$$\rho\left(v_r\frac{\partial v_z}{\partial r} + v_z\frac{\partial v_z}{\partial z}\right) = -\frac{\partial p}{\partial z} + \frac{1}{r}\frac{\partial}{\partial r}(r \tau_{rz}) + \frac{\partial \tau_{zz}}{\partial z} + \rho g, \tag{7.14b}$$

energy:
$$\rho c_p\left(v_z\frac{\partial T}{\partial z} + v_r\frac{\partial T}{\partial r}\right) = \kappa\frac{1}{r}\frac{\partial}{\partial r}\left(r\frac{\partial T}{\partial r}\right). \tag{7.15}$$

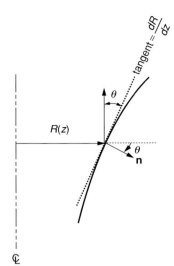

Figure 7.3. Spinline surface geometry.

Here, we have assumed that $v_\theta = 0$, that axial conduction can be neglected relative to convection, and that viscous dissipation is not important. This last assumption can easily be relaxed, as will be evident in the development, but the term does not generally seem to be significant. We have also assumed that the spinline is vertical (as it always is in commercial spinning), so gravity enters only in the axial component.

The feature that makes the analysis of this process different from those we have already studied is the presence of the free surface, which introduces a small geometrical complication in expressing the boundary conditions. The surface is shown in Figure 7.3. At the axial position z, the radius is given by $R(z)$. The outward normal vector is **n**, which is at an angle θ to the radial direction. The slope of the surface is given by

$$R' \equiv \frac{dR}{dz} = -\tan\theta. \tag{7.16}$$

Now, there is no flow across the surface, so the projection of the velocity vector **v** in the direction **n** must vanish. (In fluid mechanics terminology we would say "the surface is a streamline.") Thus,

$$r = R: \quad \mathbf{v} \cdot \mathbf{n} = v_r \cos\theta + v_z \sin\theta = 0 \tag{7.17a}$$

or, dividing by $\cos\theta$ and using Equation 7.16,

$$r = R: \quad v_r - R'v_z = 0. \tag{7.17b}$$

The total stress normal to the surface must be balanced by atmospheric pressure plus the surface tension contribution across the curved interface. We can usually neglect surface tension effects for melts, and we can always take atmospheric pressure to be zero. The tangential stress at the surface is finite because of aerodynamic drag; it is conventional to represent the stress from air drag as $\frac{1}{2}\rho_a v^2 c_D$, where ρ_a is the density of the air, v is the magnitude of the relative velocity between the surface and the air, and c_D is a dimensionless *drag coefficient* that depends on shape and on the Reynolds number, $R v \rho_a / \eta_a$, where η_a is the viscosity of the air. If the surface were cylindrical, the stress normal to the interface would simply be $\sigma_{rr} = -p + \tau_{rr}$ and the tangential stress would be τ_{rz}, but we have to deal here with the fact that $\mathbf{n}$ does not point in the r direction. Hence, there will be a small contribution from the tangential stresses to the r component and a small normal stress contribution to the z component; the resolved stress balance at the interface is thus as follows:

$$r \text{ direction:} \quad p - \tau_{rr} + R' \tau_{zr} = \tfrac{1}{2} \rho_a v^2 c_D R', \tag{7.18a}$$

$$z \text{ direction:} \quad R'(p - \tau_{zz}) + \tau_{zr} = -\tfrac{1}{2} \rho_a v^2 c_D, \tag{7.18b}$$

where v is understood to be the magnitude of the surface velocity (i.e., we are assuming that the air velocity in the spinning direction is negligible relative to the spinline speed).

Finally, the heat transfer boundary condition is the usual expression in terms of a heat transfer coefficient:

$$r = R: \quad -\kappa \frac{\partial T}{\partial r} = h(T - T_a), \tag{7.19}$$

where T_a is the ambient temperature and h is a local heat transfer coefficient. (The change in nomenclature from U to h for the heat transfer coefficient is consistent with common usage, whereby h refers to a local contribution and U to an overall coefficient that incorporates a series of heat transfer mechanisms.)

7.3.2 Thin Filament Equations

Our formal approach now follows the development that led to the equation for the cup-mixing temperature in Section 3.2.2, with two small changes. First, we are working in cylindrical coordinates, so the differential area element for integration is $2\pi r dr$. Second, we must take account of the fact that the area is changing with axial position.

Consider any dependent variable (radius, velocity, stress, temperature), which we denote $\phi(r, z)$. We define the *area average* $\overline{\phi}(z)$ as

$$\pi R^2 \overline{\phi}(z) = \int_0^R 2\pi r \phi(r, z) dr, \tag{7.20a}$$

or

$$\overline{\phi}(z) = \frac{2}{R^2} \int_0^R r \phi(r, z) dr. \tag{7.20b}$$

Recall that R is itself a function of z. To obtain the rate of change, $d\overline{\phi}/dz$, we need to use both the product rule for differentiation and Leibniz's rule for differentiating an integral:

$$\frac{d\overline{\phi}}{dz} = -\frac{4R'}{R^3} \int_0^R r\phi \, dr + \frac{2}{R^2} \int_0^R r \frac{\partial \phi}{\partial z} \, dr + \frac{2}{R} \phi \, (r, z) \, R' \qquad (7.21a)$$

or

$$\int_0^R r \frac{\partial \phi}{\partial z} \, dr = \frac{1}{2} R^2 \frac{d\overline{\phi}}{dz} + R\overline{\phi}R' - R\phi \, (R, z) \, R' = \frac{1}{2} \frac{d}{dz} \left(R^2 \overline{\phi} \right) - R\phi \, (R, z) \, R'.$$

$$(7.21b)$$

Let us illustrate the approach with the continuity equation, Equation 7.13. We multiply both terms by r (the factor 2π will simply drop out) and obtain

$$\frac{\partial}{\partial r} (rv_r) + r \frac{\partial v_z}{\partial z} = 0. \qquad (7.22)$$

Integrating from $r = 0$ to $r = R$ gives

$$rv_r \Big|_0^R + \int_0^R r \frac{\partial v_z}{\partial z} \, dr = 0 \qquad (7.23a)$$

or, using Equation 7.21b,

$$Rv_r \, (R, z) + \frac{1}{2} \frac{d}{dz} \left(R^2 \overline{v} \right) - Rv_z \, (R, z) \, R' = 0, \qquad (7.23b)$$

where we denote $\overline{v_z}$ simply by $\overline{v}$. The first and third terms sum to zero because of Equation 7.17b, which relates v_r and v_z at $r = R$, so we obtain

$$\frac{d}{dz} \left(R^2 \overline{v} \right) = 0. \qquad (7.24a)$$

$\pi R^2 \rho \overline{v}$ is the mass flow rate, which we denote w, so (with our usual assumption of a constant density) Equation 7.24a states simply that the mass flow rate is the same at every point on the spinline at steady state, which of course is what we expect:

$$w = \pi R^2 \rho \overline{v} = \text{constant}. \qquad (7.24b)$$

The details of the term-by-term integration of the r and z components of the momentum equation, Equations 7.14a and 7.14b, are carried out in Appendix 7A. The only assumption is that inertial terms can be neglected in the radial component, an assumption that is easily justified a posteriori, although the inertial terms must be retained in the axial component. We then obtain the following two averaged equations (Equations 7A.7 and 7A.10b):

$$r \text{ component:} \quad \overline{p} \, (z) = \frac{1}{2} \left(\overline{\tau}_{rr} + \overline{\tau}_{\theta\theta} \right) + \frac{1}{2} \rho_a v^2 c_D R' - \frac{1}{R^2} \frac{d}{dz} \int_0^R r^2 \tau_{rz} \, dr, \quad (7.25a)$$

$$z \text{ component:} \quad \frac{d}{dz} \left(\pi \rho R^2 \overline{v^2} \right) = \frac{d}{dz} \left[\pi R^2 \left(-\overline{p} + \overline{\tau}_{zz} \right) \right] - \pi R \rho_a v^2 c_D + \pi \rho R^2 g.$$

$$(7.25b)$$

$\overline{v^2}$ is the area average of v_z^2, and v^2 is the square of the interfacial velocity. We now assume that the deformation locally approximates a uniform, uniaxial extension, in which case we take v_z to be independent of r at each axial position, and we can neglect the contribution from the shear stress τ_{rz} relative to the normal stresses. Consistency with the assumption of a locally uniform, uniaxial extension further requires that $R' \ll 1$ ($\tan\theta \sim \theta$ in Figure 7.3), in which case v^2 will differ from $\overline{v^2}$ only by a negligible amount, and the air drag contribution to $\overline{p}$ will be negligible relative to the air drag term in the z component. Equations 7.25a and 7.25b then combine to give us

$$w\frac{d\overline{v}}{dz} = \frac{d}{dz}\left\{\pi R^2\left[\overline{\tau}_{zz} - \frac{1}{2}(\overline{\tau}_{rr} + \overline{\tau}_{\theta\theta})\right]\right\} - \pi R\rho_a\overline{v}^2 c_D + \pi\rho R^2 g. \tag{7.26}$$

Here, we made use of the fact that $\overline{v^2} \approx \overline{v}^2$ and $w = \pi\rho R^2\overline{v}$.

The details of the averaging of the energy equation, Equation 7.15, are carried out in Appendix 7A. Since we have already assumed that v_z is independent of r in deriving the final form of the momentum equation, Equation 7.26, we make the same assumption in deriving the energy equation, so the final form of the equation is expressed in terms of $\overline{T}$, rather than the cup-mixing temperature, as follows (Equation 7A.13):

$$wc_p\frac{d\overline{T}}{dz} = -2\pi Rh\left[T(R, z) - T_a\right]. \tag{7.27}$$

Equation 7.27 is reminiscent of Equation 3.25 for the extruder in that we again obtain an equation for the evolution of an *average* temperature in terms of a heat transfer term expressed through the temperature at the boundary. What has typically been done in the spinning literature has been to replace $T(R, z)$ in Equation 7.27 by $\overline{T}$ and to write the working equation as

$$wc_p\frac{d\overline{T}}{dz} = -2\pi Rh\left(\overline{T} - T_a\right). \tag{7.28}$$

The magnitude of this error has been estimated through numerical solution of the full partial differential equations, as discussed below, and the two temperatures can differ by $5\,^\circ$C under typical polyester spinning conditions.

7.3.3 Stress Constitutive Equation

The momentum equation in the form of Equation 7.26 is valid for any relation between the stress and the rate of deformation. For our purposes here we will assume that the melt can be described as an inelastic liquid and write (Table 2.3)

$$\tau_{zz} = 2\eta\frac{\partial v_z}{\partial z}, \quad \tau_{rr} = 2\eta\frac{\partial v_r}{\partial r}, \quad \tau_{\theta\theta} = 2\eta\frac{v_r}{r}. \tag{7.29}$$

We therefore have

$$\tau_{rr} + \tau_{\theta\theta} = 2\eta\left(\frac{\partial v_r}{\partial r} + \frac{v_r}{r}\right) = 2\eta\frac{1}{r}\frac{\partial}{\partial r}(rv_r) = -2\eta\frac{\partial v_z}{\partial z}. \tag{7.30}$$

Hence,

$$\tau_{zz} - \frac{1}{2}(\tau_{rr} + \tau_{\theta\theta}) = 3\eta \frac{\partial v_z}{\partial z}. \tag{7.31}$$

The averaging to obtain $\overline{\tau}_{zz} - \frac{1}{2}(\overline{\tau}_{rr} + \overline{\tau}_{\theta\theta})$ is not straightforward because η will depend on temperature (and, for a non-Newtonian fluid, on deformation rate) and will therefore be an unknown function of radial position.

Since we have already assumed that v_z is independent of r, consistency requires that we take the stretch rate $\partial v_z/\partial z$ to be independent of r and equal to $\partial \overline{v}_z/\partial z$, in which case obtaining the average of the right side of Equation 7.31 requires averaging only the viscosity; hence, we have

$$\overline{\tau}_{zz} - \frac{1}{2}(\overline{\tau}_{rr} + \overline{\tau}_{\theta\theta}) = 3\overline{\eta}\frac{\partial \overline{v}}{\partial z}. \tag{7.32}$$

Our final form for the momentum equation *for an inelastic melt* is therefore

$$w\frac{d\overline{v}}{dz} = \frac{d}{dz}\left(3\pi R^2 \overline{\eta}\frac{d\overline{v}}{dz}\right) - \pi R \rho_a \overline{v}^2 c_D + \pi \rho R^2 g \tag{7.33a}$$

or, equivalently,

$$w\frac{d\overline{v}}{dz} = \frac{d}{dz}\left(\frac{3w\overline{\eta}}{\rho\overline{v}}\frac{d\overline{v}}{dz}\right) - \left(\frac{\pi w}{\rho}\right)^{1/2} \rho_a \overline{v}^{3/2} c_D + \frac{gw}{\overline{v}}. \tag{7.33b}$$

7.3.4 Boundary Conditions

For the steady-state analysis being carried out here, the radius R can always be expressed in terms of $\overline{v}$ through the averaged form of the continuity equation, Equation 7.24b. We therefore have two dependent variables, $\overline{v}$ and $\overline{T}$. The differential equation for momentum is second order in $\overline{v}$, meaning that two constants of integration must be evaluated, while the differential equation for temperature is first order, requiring one constant of integration.

We know the temperature and mass flow rate at the spinneret; since the spinneret diameter is given, we know the melt velocity at the spinneret. Thus, we appear to know $\overline{v}(0)$ and $\overline{T}(0)$. In fact, this is not quite correct because the thin filament equations are unlikely to be a good approximation right at the spinneret, where the flow is greatly affected by the rearrangement associated with emerging from the confined spinneret flow and adjusting to the extensional free-surface deformation. Thus, the origin is typically assumed to be at an undefined position "near" the spinneret, where the thin filament equations first apply. Use of the spinneret velocity and temperature as initial conditions therefore introduces a small error, and some authors have employed correlations or simple theories for extrudate diameter adjustment ("extrudate swell" or "die swell") to obtain the initial velocity. The error introduced by using the spinneret conditions seems to be small, however, at least for PET and nylon, and well within the uncertainties introduced by other approximations in the analysis.

We know two conditions at the point of solidification, which we denote $z = L$. The solidification temperature for amorphous and slowly crystallizing polymers will typically be close to the glass transition temperature, which is known. We assume little or no deformation of the solidified fiber, so the spinline velocity at solidification will equal the takeup velocity, which is also known. Hence, for simulation purposes we may assume that we are given $\overline{v}(L)$ and $\overline{T}(L)$.[*] The boundary conditions are therefore

$$z = 0: \quad \overline{v} = v_o, \quad \overline{T} = T_o, \tag{7.34a,b}$$

$$z = L: \quad \overline{v} = v_L, \quad \overline{T} = T_L. \tag{7.34c,d}$$

We have a third-order system (three spatial derivatives), which requires evaluation of three integration constants, but we have four conditions to be satisfied. The problem is not overspecified, however, because L is in fact unknown, so we require four conditions to fix four constants, the three integration constants and the unknown length.

7.3.5 Transport Coefficients

Equations 7.26 and 7.28 contain two transport coefficients: h, the heat transfer coefficient, and c_D, the aerodynamic drag coefficient. Determining the proper values of these coefficients turns out to be a significant factor in successful modeling of the process. The simulation results are sensitive to these coefficients, and their measurement is difficult. Transport coefficients can be expected to depend on the Reynolds number of the air stream adjacent to the filament,

$$Re \equiv 2R\overline{v}\rho_a/\eta_a, \tag{7.35}$$

where the subscript a indicates that we are using the properties of air, not of the polymer. The product $R\overline{v} = w/\pi \rho \overline{v}$ is dependent on position along the spinline, so we expect h and c_D to be position dependent. Determination of the proper functional forms for h and c_D by direct measurement on an attenuating spinline is fraught with experimental uncertainty, and most measurements reported in the literature have been made on solid wires of constant diameter. Good "first principles" analyses are not available, in part because the nature of the air flow in the hydrodynamic boundary layer is not well understood.

Drag coefficient data from a number of investigators follow the form

$$c_D = \beta Re^{-0.61}, \tag{7.36}$$

---

[*] This is obviously an oversimplified picture. To handle solidification properly, it is necessary to include an equation that describes the growth of a glassy or crystalline phase and a stress equation that accounts for the biphasic nature of the system. Such descriptions are the subject of recent research, and these results should be used in new simulation codes, especially for semicrystalline polymers, but the approximation of an instantaneous transition from a liquid to an undeformable solid at a fixed average temperature $\overline{T} = T_L$ captures the important features of the spinline and suffices for our purposes here.

where the reported values of β range from 0.27 to more than 1.0; the most widely accepted value is 0.37. Heat transfer coefficient data are usually expressed in terms of the dimensionless Nusselt number,

$$Nu = 2Rh/\kappa_a, \tag{7.37}$$

where κ_a is the thermal conductivity of the air. Published simulations typically use a correlation developed by Kase and Matsuo:

$$Nu = 0.42\mathrm{Re}^{1/3}\left[1 + \left(\frac{8v_{cf}}{\bar{v}}\right)^2\right]^{1/6}, \tag{7.38}$$

where v_{cf} is the velocity of the "cross-flow" air pumped orthogonal to the filament axis. It is generally believed that this correlation overestimates heat transfer rates, and commercial fiber producers typically use proprietary heat transfer correlations in their in-house codes.[*]

7.4 Spinline Simulation

7.4.1 Temperature Development

We can obtain some useful information about spinline cooling and the distance to solidification by considering Equation 7.28, with Equation 7.38 for the heat transfer coefficient and no cross-flow air. The equation can then be written

$$\frac{d\bar{T}}{dz} = -Cw^{-5/6}\bar{v}^{1/6}\left[\bar{T} - T_a\right] \tag{7.39}$$

with $C = 1.38\kappa_a(\rho_a/\eta_a)^{1/3}/c_p\rho^{1/6}$. The dependence on $\bar{v}$ is very weak; if we assume a hundred-fold increase in spinline velocity from spinneret to solidification, for example, the coefficient $Cw^{-5/6}\bar{v}^{1/6}$ changes only by a factor of 2.2. Thus, we should be able to bound the solution by approximating $\bar{v}^{1/6}$ by a constant value and using the maximum and minimum values of $\bar{v}$ for that constant. For *fixed* $\bar{v}$ the solution of Equation 7.39 is

$$\ln\frac{\bar{T} - T_a}{T_o - T_a} = -Cw^{-5/6}\bar{v}^{1/6}z, \tag{7.40}$$

and setting $z = L$ and $\bar{T} = T_f$ we obtain

$$L = \frac{w^{5/6}\bar{v}^{-1/6}\ln\left(\frac{T_o - T_a}{T_L - T_a}\right)}{C}. \tag{7.41}$$

The minimum value of L is obtained by setting $\bar{v} = v_L$. We see that the distance to solidification is strongly dependent on the mass throughput ($w^{5/6}$) but only weakly

---

[*] Analogies between momentum and heat transfer suggest that a value of 0.60 should be used in Equation 7.36 and that the heat transfer correlation should be taken to be $Nu = 0.27\mathrm{Re}^{0.39}$ multiplied by the Kase–Matsuo cross-flow correlation factor.

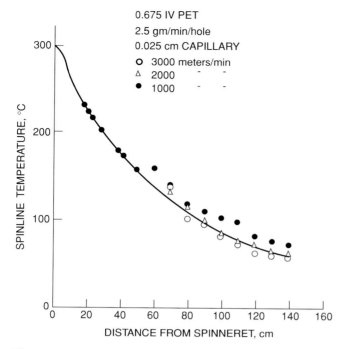

Figure 7.4. Temperature profile for PET spinning at three takeup speeds. Reprinted with permission from George, *Polym. Eng. Sci.*, **22**, 292 (1982).

dependent on the takeup or extrusion velocities. Furthermore, the temperature profile and the distance to solidification will be insensitive to the details of the fluid rheology for fixed throughput and takeup speeds.

Figure 7.4 shows the measured spinline temperatures, using a contact pyrometer, from a series of pilot plant experiments by George with a 0.675 intrinsic viscosity (*IV*) poly(ethylene terephthalate), corresponding to a molecular weight of about 45,000. The experiments are described more fully below, but it suffices here to note that throughput was held constant, while the takeup speed was varied from 1,000 to 3,000 m/min. It is clear that the profile is very insensitive to the velocity profile, as predicted by Equation 7.39. The apparent solidification point is in the midrange of the values computed from Equation 7.41 by setting $\bar{v}$ to v_o and v_L, respectively. The line through the data was computed by George from the spinline model, as discussed below.

7.4.2 Isothermal Newtonian Spinning

Commercial melt spinning processes involve substantial filament cooling, and control of the quench profile is one of the practical considerations for design and operation. Laboratory experiments are often designed to operate isothermally, however, typically by spinning in a temperature-controlled chamber with rapid solidification effected at a fixed position (by spinning into a water bath, for example) or by taking the filament up on a roller as a liquid. The viscosity is independent of temperature for isothermal spinning, and we further assume that the liquid is Newtonian, in which case η is a constant. Equation 7.33 cannot be solved analytically for constant η, but some special cases can be solved. The simplest case to consider is spinning at a low takeup speed, where inertial ($w\,d\overline{v}/dz$) and aerodynamic drag contributions can be neglected. Equation 7.33b then becomes simply

$$\frac{3w\eta}{\rho}\frac{d}{dz}\left(\frac{1}{\overline{v}}\frac{d\overline{v}}{dz}\right) = \frac{3w\eta}{\rho}\frac{d}{dz}\left(\frac{d\ln\overline{v}}{dz}\right) = 0 \tag{7.42}$$

or $d\ln\overline{v}/dz = C = $ constant, and

$$\overline{v} = v_o e^{cz}. \tag{7.43}$$

The takeup velocity v_L is specified at $z = L$, so

$$C = \ln\left(v_L/v_o\right)/L = \ln D_R/L, \tag{7.44}$$

where the *draw ratio* D_R is equal to the area reduction ratio. Hence,

$$\overline{v} = v_o e^{z\ln D_R/L}, \tag{7.45a}$$

$$R = \left(\frac{w}{\pi\rho\overline{v}}\right)^{1/2} e^{-\frac{1}{2}z\ln D_R/L}, \tag{7.45b}$$

$$\sigma_{zz} = 3\eta\frac{d\overline{v}_z}{dz} = \frac{3\eta v_o \ln D_R e^{z\ln D_R/L}}{L}. \tag{7.45c}$$

The force F on the spinline is $\pi R^2 \sigma_{zz}$:

$$F = \pi R^2 \sigma_{zz} = \frac{3\eta w \ln D_R}{\rho L}. \tag{7.45d}$$

The force is independent of position on the spinline, as we expect; Equation 7.33a is simply $dF/dz = 0$ when we neglect inertia, aerodynamic drag, and gravity. The solution obtained here is useful for analyzing process instabilities, where the essential features are often present even without the phenomena we have neglected. We address spinline instabilities in Chapter 11.

7.4.3 Numerical Solution

Numerical solution of Equations 7.28 and 7.33 is straightforward. We are given v_o and T_o. We assume a value of $d\overline{v}/dz$ at $z = 0$ (equivalent to assuming the initial stress). These three conditions define a well-posed initial value problem and are sufficient to integrate the equations numerically, using any algorithm. The equations are integrated until $T = T_L$, defining the length L. If $v = v_f$ at $z = L$, we have a solution to the boundary value problem; if not, we change the assumed initial condition $d\overline{v}/dz$ and repeat. Alternatively, we can use the condition $v = v_f$ to define L and check to see if $T = T_L$, or we can use the first of the two conditions at $z = L$ to be satisfied. The process involves only a one-dimensional search for the proper

Table 7.1. *Conditions for George's spinning studies on 0.675 IV PET*

Extrusion temperature	295 °C
Quench air temperature	30 °C
Air cross-flow velocity	0.2 m/s
Spinneret hole diameter	0.254 mm (10 mil)
Throughput/spinneret hole	2.5 g/min
Spinneret velocity	18.2 m/min (0.3 m/s)

initial value of $d\bar{v}/dz$, and the stress–takeup velocity relation is monotonic, so convergence is usually rapid. The search can be carried out efficiently, for example, using a Newton-Raphson scheme. There is no difficulty in principle in including a position dependence of the cross-flow air velocity and of the physical properties of the air.

Nondimensionalization of the equations, which is usually desirable, can lead in this case to a computational complication. Many authors have used the length L to nondimensionalize axial position so that the dimensionless length varies from 0 to 1. In that case L enters the problem formulation explicitly, and both the length and initial stress must be adjusted on each iteration, requiring a two-variable search. Convergence of the iterations is still rapid, in part because Equation 7.41 gives an excellent initial approximation for L, but clearly such a formulation should be avoided.

7.4.4 Simulation of PET Pilot Plant Data

A number of authors have reported experimental spinning data and simulations. Early work by Kase and co-workers at Toyobo was very influential, but we prefer to use an excellent data set by Henry George of Celanese, who studied the spinning of 0.675 intrinsic viscosity poly(ethylene terephthalate) fibers on a pilot-scale spinning machine under commercial conditions. (We have already seen some of these data in Figure 7.4.) The spinning conditions are given in Table 7.1. Solidification was taken to occur at 70 °C in the simulations, which is approximately the glass transition temperature of PET. We show results here from simulations carried out by Gagon and Denn and by George, with the assumption that PET is a Newtonian liquid. (PET is in fact slightly viscoelastic, although the viscosity is insensitive to deformation rate, and we will return to this example in Chapter 10.)

The viscosity of PET was reported by Gregory in 1973 as

$$\eta = 1.13 \times 10^{-14} M w^{3.5} \exp\left(-11.98 + 6800/T\right) \tag{7.46}$$

with η in pascal seconds and T in Kelvins. Gregory's data were reported at 265 °C and above, and it was assumed in the simulations that the same temperature dependence applied all the way to 70 °C. In fact, a different temperature dependence (the Williams-Landel-Ferry [WLF] equation) should be used within 100 °C of the glass transition temperature; this change is incorporated in a computer code by Kase and commercialized by Toyobo, but the difference in computed behavior is not

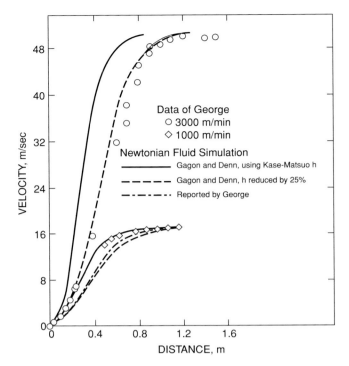

Figure 7.5. Simulation of PET spinline data. Reprinted from Gagon and Denn, *Polym. Eng. Sci.*, **21**, 844 (1981).

significant. All physical properties were taken as temperature dependent, with the relations obtained from standard sources. Air properties were evaluated at the mean of the polymer and ambient temperatures to reflect the heating of the air boundary layer.

Gagon and Denn used the Kase–Matsuo heat transfer correlation (Equation 7.38) as a base case, while George used a correlation derived from in-house Celanese data. George's correlation tends to give higher heat transfer rates than Kase and Matsuo's. (There is a suggestion in the text that George averaged two calculations with different quench air temperatures to account for the different environments experienced by filaments in different locations, but this is not clear.) Both studies employed Equation 7.36 for the drag coefficient; Gagon and Denn took $\beta = 0.37$, while George took $\beta = 0.44$ but apparently included a correction factor to account for reduced drag because of the presence of multiple filaments. The initial filament radius was estimated in both cases from a Celanese correlation for extrudate swell that predicted an area increase of 100%, so v_o was taken to be 9.1 m/min.

Simulation results are shown together with velocity profiles in Figure 7.5 for takeup speeds of 1,000 and 3,000 m/min. Gagon and Denn's calculations with the Kase–Matsuo heat transfer coefficient overestimate the rate of velocity increase for the 3,000 m/min data, but they are in reasonable agreement at 1,000 m/min. There is a substantial improvement in the fit to the higher takeup speed data when the

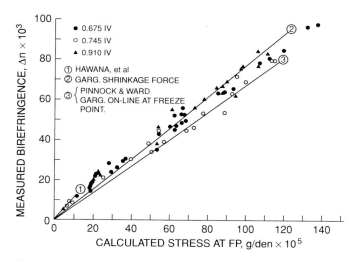

Figure 7.6. Measured birefringence as a function of calculated takeup stress for three PET melts. Reprinted with permission from George, *Poly. Eng. Sci.*, **22**, 292 (1982).

coefficient in Equation 7.38 is reduced by 25%, from 0.42 to 0.315, which is within the uncertainty of most heat transfer data, but at the expense of a somewhat poorer fit at the lower takeup speed. George's calculations are close to the latter curves, suggesting that the actual heat transfer rate used in his simulations is less than that given by his reported correlation, perhaps because of a multifilament correction. The sensitivity to the choice of β in Equation 7.36 is considerably less, and while the fit to data is better with $\beta = 0.37$, one could not rule out even a value of $\beta = 0.60$ based on this data set.

The two simulations, which differ only in minor detail, indicate that velocity and temperature profiles for PET can be computed with sufficient confidence to investigate trends, and *this is the primary use of computer codes implementing these and similar models*. We can ask "what if" questions about operating conditions, physical properties and so forth, and use the model to guide experiment.

Many important physical properties of amorphous yarns (e.g., tenacity, elongation to break, modulus) correlate with the *optical birefringence*, which is the difference in indices of refraction in orthogonal directions. The birefringence is caused by orientation of the polymer chains, which is also the cause of the frozen stress in the filament. Simple molecular theories of polymer melt mechanics predict proportionality between stress and birefringence, and this is observed experimentally; indeed, knowledge of the proportionality factor, or the *stress–optical coefficient*, enables the use of nonintrusive methods to measure stresses. One potential use of a spinline model thus becomes obvious: If we can predict the stress at solidification, we can, with a knowledge of the stress–optical coefficient, predict the birefringence; hence, we can predict the properties that correlate with birefringence.

Figure 7.6 shows a plot by George of *measured* birefringence plotted versus *calculated* stress at the solidification point for a range of takeup speeds and three different PET resins, the 0.675 *IV* resin used for the previous simulations and two of higher molecular weight. (g/denier is an industry unit equal to grams/9,000 meters of length.) There is scatter, but the points fall near a straight line, indicating a constant stress–optical coefficient. The straight line labeled 1 is from a similar modeling study limited to low stresses, whereas Lines 2 and 3 are the results of experimental studies where the stress was measured by a variety of methods. The good agreement is a further indication of the usefulness of the spinline model.

7.4.5 Radial Temperature Variation

Radial temperature variation across the filament can be important. There are patents, for example, based on *differential* birefringence, the radial variation of optical birefringence on the drawn filament; from the modeling point of view, the radial variation in birefringence is determined by the radial variation in the tensile stress, which results from a temperature-induced radial variation in the viscosity. There is an elementary and seemingly adequate approach to finding the radial variation, in which we continue to assume that the kinematics are locally extensional, so $v_r(r, z) = -1/2 r \, d\bar{v}/dz$ (cf. Equation 7.3). We continue to solve the thin filament Equation 7.33 for $\bar{v}(z)$, but we solve the full partial differential Equation 7.15 for $T(r, z)$. We compute $\bar{\eta}(z)$ from the radial temperature profile at each step of the integration. The stress (hence, birefringence) distribution is calculated from

$$\sigma_{zz}(r, z) = 3\eta \left(T\left(r, z\right)\right) \frac{d\bar{v}\left(z\right)}{dz}. \tag{7.47}$$

The computational algorithm required is more complex than the one used to solve the strictly one-dimensional model, but it employs straightforward methodologies. Figure 7.7 shows a calculation of the temperature by Vassilatos and co-workers using parameters similar to those in the preceding example, but with $T_o = 290\ ^\circ\text{C}$, $w = 4.0\ \text{g/min}$, $v_L = 3,500\ \text{m/min}$, and no cross-flow air. There is about a 10° temperature difference from the surface (T_R) to the centerline (T_C).

   It is possible to obtain an analytical estimate of the temperature profile in the lower portion of the melt zone, which is the region of interest in terms of drawn-filament properties. If we change the independent variables from (r, z) to (ξ, z), where $\xi = r/R(z)$, Equation 7.15 becomes

$$\frac{\partial T}{\partial z} = \frac{\pi \kappa}{w c_p} \frac{1}{\xi} \frac{\partial}{\partial \xi} \xi \frac{\partial T}{\partial \xi} \tag{7.48}$$

with boundary conditions

$$\frac{\partial T}{\partial \xi} = 0 \quad \text{at} \quad \xi = 0, \frac{\partial T}{\partial \xi} = -Bi \left(T - T_a\right) \text{ at } \xi = 1. \tag{7.49a,b}$$

$Bi = hR/\kappa$ is a Biot number, where κ is the thermal conductivity of the polymer; the Biot number is related to the Nusselt number by

$$2Bi = \frac{\kappa_a}{\kappa} Nu. \tag{7.50}$$

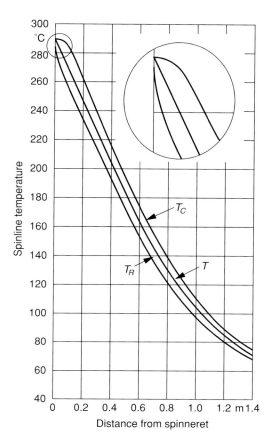

Figure 7.7. Spinline radial temperature variation. Reprinted from Vassilatos et al., *Int. Polym. Proc.*, **VII**, 144 (1992).

Equation 7.48 with boundary conditions 7.49 is simply the equation for transient conductive heat transfer in a cylinder, where z is the timelike variable, except that Bi varies with z. We know that Bi is relatively insensitive to the velocity and can be approximated by a constant value in the lower portion of the spinline, where little attenuation occurs. The solution to Equations 7.48 and 7.49 for *constant Bi* is

$$T - T_a = \sum_{n=1}^{\infty} C_n J_o \left(\lambda_n r/R\right) \exp\left(-\lambda_n^2 \pi \kappa z/w c_p\right). \tag{7.51}$$

J_o is the *Bessel function* of the first kind of zero order (a tabulated function); the $\{\lambda_n\}$ are roots of

$$\lambda J_1\left(\lambda\right) = Bi J_o\left(\lambda\right), \tag{7.52}$$

where J_1 is the Bessel function of the first kind of first order. These roots are tabulated. The constants $\{C_n\}$ are determined from known conditions at some value of z, using the orthogonality property of the Bessel functions.

We can obtain the temperature distribution near the solidification region without solving for the constants $\{C_n\}$. The first term in the series for T dominates when $\lambda_2^2 \pi \kappa z/w c_p \geq 3$, since $\exp(-3)$ is essentially zero; λ_2 is close to 4 over a large range of Bi, so this inequality will be satisfied even in the upper portion of the spinline. In that case we have

$$T \cong T_a + C_1 J_o \left(\lambda_1 r / R \right) \exp \left(-\lambda_1^2 \pi \kappa z / w c_p \right) \tag{7.53a}$$

and

$$\overline{T} \cong T_a + C_1 \left[\frac{2}{R^2} \int_0^R r J_o \left(\lambda_1 r / R \right) dr \right] \exp \left(-\lambda_1^2 \pi \kappa z / w c_p \right)$$

$$= T_a + \frac{2C_1}{\lambda_1} J_1(\lambda_1) \exp \left(-\lambda_1^2 \pi \kappa z / w c_p \right). \tag{7.53b}$$

Thus, with Equation 7.51, we obtain

$$\frac{T - T_a}{\overline{T} - T_a} \cong \left(\frac{\lambda_1^2}{2 Bi} \right) \frac{J_o \left(\lambda_1 r / R \right)}{J_o \left(\lambda_1 \right)}. \tag{7.54}$$

Values of $Nu \sim 2$ are typical. κ for PET is about 0.19 J/m.s.K, while κ_a for air is about 0.03. A typical value for Bi would thus be of order 0.1, and for such small values of Bi we have $2 Bi \sim \lambda_1^2$. Thus, setting $r = R$ in Equation 7.52 we find a difference of about $7\,^{\circ}$C between the surface and average temperatures when $\overline{T} = 70\,^{\circ}$C and $T_a = 30\,^{\circ}$C.

7.5 Outstanding Issues

The spinline model described here has a number of important limitations. The first, of course, is the assumption of an inelastic description for the melt rheology. This description is adequate for PET, and probably for nylon, but it will be inadequate for polypropylene. We will return to this issue subsequently, after we discuss the relevant rheology in Chapter 9.

Next, we are generally concerned with multifilament spinning, in which large numbers of filaments extruded from a single spinning head are gathered together following solidification to form a yarn. The aerodynamic drag and heat transfer coefficients for the individual filaments must be adjusted to allow for the effect of filament–filament interaction on the aerodynamic boundary layer. A number of ways of dealing with the multifilament problem have been proposed and implemented, but a truly satisfying solution has not been described in the literature.

We have said little here about the proper selection of initial conditions for the thin filament equations. This issue has received some attention through the use of numerical methods like those described in the next chapter, which make it possible to link the shear and exit flow in the spinneret to the spinline. A related but more difficult problem that has received very little attention is the proper modeling of heat transfer at the top of the spinline, where crossflow air is likely to be impeded by the geometry, and heat transfer from the metal spinneret plate might be important.

Finally, the assumption that solidification occurs instantaneously at a specified temperature seems to be adequate for an amorphous polymer like PET, but it is clearly inadequate when there is significant crystallinity, as there will be with polypropylene, nylon, and even PET at takeup speeds well above 5,000 m/min. It is not difficult in principle to include crystallinity in a model like the one developed here: We need a kinetic equation for the rate of crystallization, a term in the energy equation to account for heat effects associated with crystallization, and modification of the stress term to account for the portion of the stress carried by the solid phase. Stress-induced crystallization is poorly understood, but recent results by Doufas and McHugh using the approach outlined here show good agreement with pilot-scale data for nylon and high-speed spinning of PET. The latter work is based on a viscoelastic model of spinning, which we address in Chapter 10.

BIBLIOGRAPHIC NOTES

Some of the material in this chapter is covered in the texts by Middleman, Pearson, and Tadmor and Gogos cited previously. The standard text on the mechanics of fiber spinning, which predates most of the published work on spinline simulation, is

Ziabicki, A., *Fundamentals of Fibre Spinning*, John Wiley, New York, 1976.

The present chapter largely follows the treatments in

Denn, M. M., "Fibre Spinning," in J. R. A. Pearson and S. M. Richardson, Eds., *Computational Analysis of Polymer Processing*, Applied Science, London, 1983, pp. 179ff.

Vassilatos, G., E. R. Schmelzer, and M. M. Denn, *Int. Polym. Proc.*, **VII**, 144 (1992).

Correlations for the transport coefficients are addressed in

Denn, M. M., *Ind. Eng. Chem. Res.*, **35**, 2842 (1996).

Miller, C., *AIChE J.*, **50**, 898 (2004).

The effect of aerodynamics on drag and heat transfer in multifilament spinning is studied using a computational fluid dynamics code for the air flow in

Harvey, A. D., and A. K. Doufas, *AIChE J.*, **53**, 78 (2007).

The spinline model used by Harvey and Doufas assumes that the melt is viscoelastic and crystallizable. We will discuss viscoelastic models in Chapter 10.

Appendix 7.A Thin Filament Equations

The z momentum equation is

$$\rho v_r \frac{\partial v_z}{\partial r} + \rho v_z \frac{\partial v_z}{\partial z} = -\frac{\partial p}{\partial z} + \frac{1}{r}\frac{\partial}{\partial r}(r\tau_{rz}) + \frac{\partial \tau_{zz}}{\partial z} + \rho g. \qquad (7.14b)$$
$$\quad\text{(a)}\qquad\text{(b)}\qquad\quad\text{(c)}\qquad\text{(d)}\qquad\text{(e)}\qquad\text{(f)}$$

We multiply each term by rdr and integrate. The first term on the left becomes

$$\int_0^R r\rho v_r \frac{\partial v_z}{\partial r}dr = \rho r v_r v_z\Big|_0^R - \int_0^R \rho v_z \frac{\partial}{\partial r}(rv_r)\,dr = \rho r v_r v_z\Big|_0^R + \int_0^R \rho r v_z \frac{\partial v_z}{\partial z}dr, \quad (7A.1)$$

where the last substitution has employed the continuity equation, Equation 7.13. We thus have two identical terms, $\int_0^R r\rho v_z \frac{\partial v_z}{\partial z}dr$, one in (a) and one in (b). Turning to term (b) on the left,

$$\int_0^R r\rho v_z \frac{\partial v_z}{\partial z} dr = \frac{1}{2}\rho \int_0^R r\frac{\partial v_z^2}{\partial z} dr = \frac{1}{4}\rho\frac{d}{dz}\left(R^2\overline{v_z^2}\right) - \frac{1}{2}\rho R v_z^2(R,z) R'. \qquad (7A.2)$$

Summing the terms in Equation 7A.1 and 7A.2, we therefore obtain

$$(a) + (b) = \frac{1}{2}\rho\frac{d}{dz}\left(R^2\overline{v_z^2}\right) + \rho R v_z(R,z)\left[v_r(R,z) - v_z(R,z) R'\right]$$

$$= \frac{1}{2}\frac{d}{dz}\left(R^2\overline{v_z^2}\right), \qquad (7A.3)$$

where we recall again that the integral appears twice, and we have used Equation 7.17b to relate v_r and v_z at $r = R$. Note that the only remaining term in Equation 7A.3 involves the average of v_z^2, not of v_z.

Turning now to the right-hand side of Equation 7.14b, we combine terms (c) and (e) to write

$$\int_0^R r\frac{\partial}{\partial z}(-p + \tau_{zz}) dr = \frac{1}{2}\frac{d}{dz}\left[R^2(-\overline{p} + \overline{\tau}_{zz})\right] - R\left[-p(R,z) + \tau_{zz}(R,z)\right] R'.$$

$$(7A.4)$$

Term (d) becomes

$$\int_0^R \frac{\partial}{\partial r}(r\tau_{rr}) dr = R\tau_{rr}(R,z). \qquad (7A.5)$$

Combining these two expressions, and noting that the ρg term integrates simply to $\frac{1}{2}R^2\rho g$, we obtain

$$(c) + (d) + (e) = \frac{1}{2}\frac{d}{dz}\left[R^2(-\overline{p} + \overline{\tau}_{zz})\right] - \frac{1}{2}\rho_a v^2 c_D R + \frac{1}{2}\rho R^2 g, \qquad (7A.6)$$

where we have used Equation 7.18b for the stress at $r = R$. v is the magnitude of the relative velocity between the filament and the air at $r = R$. Multiplying each term by 2π and combining thus gives us

$$\frac{d}{dz}\left(\pi\rho R^2\overline{v_z^2}\right) = \frac{d}{dz}\left[\pi R^2(-\overline{p} + \overline{\tau}_{zz})\right] - \pi\rho_a v^2 R c_D + \pi\rho R^2 g. \qquad (7A.7)$$

The averaging of the r component, Equation 7.14a, is more delicate. First, we neglect the inertial terms in the radial direction. Second, after multiplying the remaining terms by r and integrating, we find that in place of the desired forms $\int r p \, dr, \int r\tau_{\theta\theta} dr$, and so forth, we obtain integrals of the form $\int p \, dr, \int \tau_{\theta\theta} dr$. This leads us to conclude that the proper way to average the terms is to multiply by r^2 and integrate; that is,

$$0 = -\int_0^R r^2\frac{\partial p}{\partial r} dr + \int_0^R r\frac{\partial}{\partial r}(r\tau_{rr}) \, dr - \int_0^R r\tau_{\theta\theta} dr + \int_0^R r^2\frac{\partial\tau_{rz}}{\partial z} dr. \qquad (7A.8)$$

$$\quad\quad\quad (a) \quad\quad\quad\quad\quad (b) \quad\quad\quad\quad (c) \quad\quad\quad (d)$$

The integrations are carried out as follows:

$$(a): -\int_0^R r^2 \frac{\partial p}{\partial r} dr = -R^2 p\,(R) + 2\int_0^R r p\, dr = -R^2 p\,(R) + R^2 \overline{p}, \qquad (7A.9a)$$

$$(b): \int_0^R r \frac{\partial}{\partial r}(r\tau_{rr})\,dr = R^2 \tau_{rr}\,(R) - \int_0^R r\tau_{rr}\,dr = R^2 \tau_{rr}\,(R) - \frac{1}{2}R^2 \overline{\tau}_{rr}, \qquad (7A.9b)$$

$$(c): -\int_0^R r\tau_{\theta\theta}\,dr = -\frac{1}{2}R^2 \overline{\tau}_{\theta\theta}, \qquad (7A.9c)$$

$$(d): \int_0^R r^2 \frac{\partial \tau_{rz}}{\partial z}\,dr = \frac{d}{dz}\int_0^R r^2 \tau_{rz}\,dr - R^2 \tau_{rz}\,(R)\,R'. \qquad (7A.9d)$$

Summing the four terms, we obtain

$$R^2\left[\overline{p} - \frac{1}{2}(\overline{\tau}_{rr} + \overline{\tau}_{\theta\theta}) - p\,(R) + \tau_{rr}\,(R) - \tau_{rz}\,(R)\,R'\right] + \frac{d}{dz}\int_0^R r^2 \tau_{rz}\,dr = 0$$

$$(7A.10a)$$

or, using Equation 7.18a for the stresses at the boundary,

$$\overline{p} = \frac{1}{2}(\overline{\tau}_{rr} + \overline{\tau}_{\theta\theta}) + \frac{1}{2}\rho_a v^2 c_D R' - \frac{1}{R^2}\frac{d}{dz}\int_0^R r^2 \tau_{rz}\,dr. \qquad (7A.10b)$$

To average the energy equation, Equation 7.15, we again multiply by r and integrate as follows:

$$\rho c_p \int_0^R r v_z \frac{\partial T}{\partial z}\,dr + \rho c_p \int_0^R r v_r \frac{\partial T}{\partial r}\,dr = \kappa \int_0^R \frac{\partial}{\partial r}\left(r\frac{\partial T}{\partial r}\right)dr. \qquad (7A.11)$$

$$(a) \qquad\qquad (b) \qquad\qquad (c)$$

Looking at the second term on the left, we have

$$\int_0^R r v_r \frac{\partial T}{\partial r}\,dr = R v_r(R)\,T(R) - \int_0^R T\frac{\partial}{\partial r}(r v_r)\,dr = R v_r(R)\,T(R) + \int_0^R r T\frac{\partial v_z}{\partial z}\,dr.$$

$$(7A.12a)$$

Noting that term (a) contains $v_z \partial T/\partial z$ and term (b) contains $T\partial v_z/\partial z$, we combine the two using the product rule for derivatives to a single term $\partial(T v_z)/\partial z$ and write

$$(a) + (b): \rho c_p\left[R v_r(R)\,T(R) + \int_0^R r\frac{\partial (v_z T)}{\partial z}\,dr\right]$$

$$= \rho c_p\left[R v_r(R)\,T(R) + \frac{1}{2}\frac{d}{dz}(R^2 \overline{v_z T}) - R v_z(R)\,T(R)\,R'\right]. \qquad (7A.12b)$$

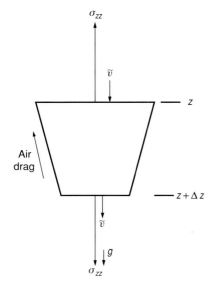

Figure 7B.1. Free-body diagram of a spinline segment.

The first and third terms sum to zero because of the kinematic boundary condition Equation 7.17b. We have already assumed that v_z is independent of r, in which case v_z can be removed from the integral in computing $\overline{v_z T}$, and we have $\overline{v_z T} = \overline{v_z}\,\overline{T}$. Hence, the remaining term in Equation 7A.12b can be written as $wc_p\left(d\overline{T}/dz\right)/2\pi$.

Integration of the right-hand side of Equation 7A.11 gives

$$\kappa \int_0^R \frac{\partial}{\partial r}\left(r\frac{\partial T}{\partial r}\right)dr = \kappa R \frac{\partial T}{\partial r}\bigg|_{r=R} = -Rh\left[T\left(R\right) - T_a\right], \qquad (7A.12c)$$

where the final equality follows from the boundary condition Equation 7.19. Equating the terms in Equation 7A.12b and 7A.12c then gives

$$wc_p \frac{d\overline{T}}{dz} = -2\pi Rh\left[T\left(R\right) - T_a\right]. \qquad (7A.13)$$

APPENDIX 7.B Macroscopic Derivation

Consider the control volume, or "free body," shown in Figure 7B.1. Conservation of mass states that the rate of change of mass in the control volume equals the rate at which mass flows in at surface z less the rate at which mass flows out at surface $z + \Delta z$. The mass flow rate is $\pi R^2 \rho \overline{v}$, while the total mass in the control volume is $\pi R^2 \pi \Delta z$; thus,

$$\frac{\partial}{\partial t}\pi R^2 \rho \Delta z = \pi R^2 \rho \left.\overline{v}\right|_z - \pi R^2 \rho \left.\overline{v}\right|_{z+\Delta z}. \qquad (7B.1)$$

The time derivative is a partial derivative because position is being held constant. Dividing by Δz and taking the limit as $\Delta z \to 0$ then gives

$$\frac{\partial}{\partial t} \pi R^2 \rho = -\frac{\partial}{\partial z} \pi R^2 \rho \overline{v} \tag{7B.2a}$$

or, since ρ is assumed to be a constant,

$$\frac{\partial R^2}{\partial t} = -\frac{\partial}{\partial z} R^2 \overline{v}. \tag{7B.2b}$$

At steady state $\partial R^2 / \partial t = 0$, so $\pi R^2 \rho \overline{v} = w$ is a constant, independent of z.

Conservation of linear momentum states that the rate of change of linear momentum in the control volume equals the net flow rate of linear momentum in and out, plus the sum of the imposed forces. Momentum per unit mass is the velocity; thus, the total z momentum in the control volume is $(\pi R^2 \rho \Delta z)\overline{v}$, and the flow rate of z momentum is $(\pi R^2 \rho \overline{v})\overline{v}$. The tensile stress imposed by the surrounding fluid is $\overline{\sigma}_{zz}$; the direction for this stress as shown in Figure 7B.1 represents the usual convention in mechanics, in which a positive stress puts the control volume in tension (i.e., a pressure is negative stress). Air drag, which is written as the product of an impact pressure $(\frac{1}{2}\rho_a \overline{v}^2)$, the surface area over which the stress operates $(2\pi R \Delta z)$, and a drag coefficient c_D, acts opposite to the direction of motion. We assume the spinning direction is in the direction of gravity. Surface tension forces can usually be neglected for melt spinlines. Thus, we obtain

$$\frac{\partial}{\partial t} \pi R^2 \rho \overline{v} \Delta z = \pi R^2 \rho \left. \overline{v}^2 \right|_{z+\Delta z} - \pi R^2 \rho \left. \overline{v}^2 \right|_z + \pi R^2 \left. \overline{\sigma}_{zz} \right|_{z+\Delta z} - \pi R^2 \left. \overline{\sigma}_{zz} \right|_z$$

$$- \left(\frac{1}{2} \rho_a \overline{v}^2 \right) (2\pi R \Delta z) c_D + \pi R^2 \Delta z g. \tag{7B.3}$$

Dividing by Δz and taking the limit $\Delta z \to 0$ gives

$$\frac{\partial}{\partial t} \pi R^2 \overline{v} = -\frac{\partial}{\partial z} \pi R^2 \rho \overline{v}^2 + \frac{\partial}{\partial z} \pi R^2 \overline{\sigma}_{zz} - \rho_a \pi R \overline{v}^2 c_D + \pi R^2 \rho g. \tag{7B.4a}$$

This equation is simplified somewhat by expanding derivatives of products and using Equation 7B.2 to obtain

$$\pi R^2 \rho \frac{\partial \overline{v}}{\partial t} = -w \frac{\partial \overline{v}}{\partial z} + \frac{\partial}{\partial z} \pi R^2 \overline{\sigma}_{zz} - \rho_a \pi R \overline{v}^2 c_D + \pi R^2 \rho g. \tag{7B.4b}$$

Conservation of energy states that the rate of change of the total energy in the control volume equals the net rate of energy flow in and out, plus the net rate at which work is done on the control volume by the surroundings, plus the rate of heat flow into the control volume. Kinetic and potential energy changes are usually small relative to internal energy changes, so only internal energy needs to be considered. Total internal energy in the control volume is $\underline{E}\pi R^2 \rho \Delta z$, where $\underline{E}$ is the internal energy per unit mass. The work/unit mass done to move fluid into and out of the control volume,* p/ρ, is combined with the internal energy flow term to give the flow of *enthalpy* per unit mass, $\underline{H} = \underline{E} + p/\rho$. We then have

$$\frac{\partial}{\partial t}\underline{\overline{E}}\pi R^2\rho\Delta z = \pi R^2\rho\overline{v}\,\underline{\overline{H}}\Big|_z - \pi R^2\rho\overline{v}\,\underline{\overline{H}}\Big|_{z+\Delta z} - 2\pi Rh\Delta z\left(\overline{T} - T_a\right), \qquad (7B.5)$$

where h is the heat transfer coefficient. We have neglected work done on the control volume by the extra stresses. Dividing by Δz and taking the limit as $\Delta z \to 0$ gives

$$\frac{\partial}{\partial t}\underline{\overline{E}}\pi R^2\rho = -\frac{\partial}{\partial z}\pi R^2\rho\overline{v}\underline{\overline{H}} - 2\pi Rh\left(\overline{T} - T_a\right) \qquad (7B.6a)$$

or, using Equation 7B.2,

$$\pi R^2\rho\frac{\partial\underline{\overline{E}}}{\partial t} = -w\frac{\partial\underline{\overline{H}}}{\partial z} - 2\pi Rh\left(\overline{T} - T_a\right). \qquad (7B.6b)$$

If $\underline{E}$ and $\underline{H}$ are taken to depend only on temperature, then $\partial\underline{E} = c_v\partial T$ and $\partial\underline{H} = c_p\partial T$, where c_v and c_p are the heat capacities at constant volume and constant pressure, respectively. For liquids, c_p and c_v are numerically almost equal. Thus, we obtain

$$\pi R^2\rho c_p\frac{\partial\overline{T}}{\partial t} = -wc_p\frac{\partial\overline{T}}{\partial z} - 2\pi Rh\left(\overline{T} - T_a\right). \qquad (7B.7)$$

---

* The flow work should include the full normal stress, $\sigma_{zz} = -p + \tau_{zz}$, rather than just the isotropic pressure. The derivation then requires some additional steps that lead to Equation 7B.7, but with the viscous dissipation term included.

8 Numerical Simulation*

8.1 Introduction

The examples we have studied thus far have had rather simple kinematics: flow parallel or nearly parallel to a wall and ideal or nearly ideal extension. Thus, we have been able to obtain exact solutions for the flow or to obtain approximate solutions based on the small difference between the actual flow and an ideal case for which an exact solution is available. Even for the case of fiber spinning, where an analytical solution to the thin filament equations cannot be obtained under conditions relevant to industrial practice, we simply need to obtain a numerical solution to a pair of ordinary differential equations, which is a task that can be accomplished using elementary and readily available commercial software.

The flow in many real processing geometries is too complex for us to apply the analytical methods utilized in the preceding chapters. Indeed, even when the flow field is a simple one, the coupled heat transfer problem may not be amenable to a simple treatment; the elementary extruder in Chapter 3 is an example of a case in which we are unable to obtain an exact or even approximate solution for the spatial development of the two-dimensional temperature field.

Complex coupled flow and heat transfer problems can be solved using numerical techniques in which the partial differential equations are converted to a large set of coupled algebraic equations, and the algebraic equations are then solved using conventional methods developed specifically to be efficient on digital computers. The concept by which the numerical solution of the partial differential equations is obtained is rather straightforward, and we will describe it here. Actual implementation into an efficient, user-friendly computer code is difficult and tedious, however, and most users employ commercial software.

8.2 Galerkin's Method

Most modern computer codes for low Reynolds number flow, the case in which we are generally interested in polymer processing applications, use an approximation

---

* This chapter was coauthored by Benoit Debbaut, ANSYS/Polyflow s.a., Belgium.

technique known as *Galerkin's method*, which is one of a group of methods known collectively as *methods of weighted residuals*. The general idea behind all methods of weighted residuals is the same: Assume a shape (i.e., a functional form) for the solution to the differential equation, but with one or more unknown coefficients that must be specified; we call this the *approximate solution*. The approximate solution is then put into the differential equation. It will not satisfy the differential equation everywhere because it is an approximate solution, so there is an error – that is, something is left over – at each point in space; we call this error the *residual*, and the residual will depend on the set of unknown coefficients, which we denote $\{C_n\}$. We then seek to distribute the error by setting a weighted average of the residual to zero. In fact, we set as many weighted averages to zero as we have unknown coefficients $\{C_n\}$, thus obtaining a set of *algebraic* equations for the unknown coefficients.

We must therefore select the set of functions $\{\varphi_n\}$ that we will use to approximate the solution and the set of weighting functions $\{w_n\}$ that we will use to compute the weighted averages of the residual. Galerkin's method uses the same set of functions for both purposes. This is an unexpected and seemingly curious choice, but it has a rigorous foundation for certain classes of problems in a branch of mathematics known as the calculus of variations, which is the study of maxima and minima. Possibly the most familiar connection between a minimization problem and a problem in mechanics is linear elasticity, where the stress field over the interior of an object resulting from forces imposed at the boundary minimizes the total energy. The creeping flow equations have a mathematical structure that is analogous to linear elasticity. In general, all differential equations that are *self-adjoint*, a class that includes many of the equations of mathematical physics, correspond to minimization problems in the calculus of variations.

As an example of the application of Galerkin's method, consider the following linear boundary-value problem:

$$y'' + y + x = 0, \quad y(0) = y(1) = 0. \tag{8.1}$$

This equation has an analytical solution $y = (\sin x / \sin 1) - x$. Now suppose we wish to seek an approximate solution in the form

$$y = \sum_n C_n x^n (1 - x); \tag{8.2}$$

that is, the approximating functions φ_n are of the form $x^n(1 - x), n = 1, 2, \ldots, N$. The residual $\mathfrak{R}$ is then

$$\mathfrak{R}(\{C_n\}; x) = \sum_{n=2} C_n n(n - 1) x^{n-2} - \sum_{n=1} C_n n(n + 1) x^{n-1} + \sum_{n=1} C_n x^n (1 - x) - x. \tag{8.3}$$

We now multiply the residual by each of the approximating functions $x^m(1 - x)$ in turn for $m = 1$ to N, integrate from $x = 0$ to $x = 1$, and set the integral to zero:

$$\int_0^1 \mathfrak{R}(\{C_n\}; x) x^m (1 - x) dx = 0, \quad m = 1, 2, \ldots, N. \tag{8.4}$$

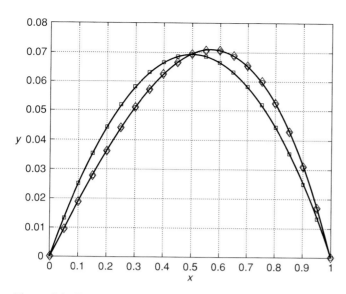

Figure 8.1. Exact solution ($\circ$) and one- ($\square$) and two- ($\diamond$) term Galerkin approximations to Equation 8.1.

Equation 8.4 is a set on N linear algebraic equations for the N constants $\{C_n\}$.
   Suppose we decide that a one-term approximation is sufficient. We then set $N = 1$ and obtain

$$\left[\int_0^1 [-2 + x(1-x)x(1-x)]dx\right] C_1 - \int_0^1 x^2(1-x)dx = 0, \qquad (8.5)$$

and $C_1 = 5/18 = 0.2777\ldots$, or

$$y \approx 0.278x(1-x). \qquad (8.6)$$

We obtain a higher order polynomial approximation by retaining two terms in the expansion, in which case we obtain two coupled linear equations for the constants C_1 and C_2:

$$\frac{3}{10}C_1 + \frac{3}{20}C_2 = \frac{1}{12}, \qquad (8.7a)$$

$$\frac{3}{20}C_1 + \frac{13}{105}C_2 = \frac{1}{20}. \qquad (8.7b)$$

We then obtain $C_1 = 0.192$ and $C_2 = 0.171$; hence,

$$y \approx 0.192x(1-x) + 0.171x^2(1-x). \qquad (8.8)$$

The symmetry of the matrix of coefficients in Equations 8.7a–b is a consequence of the self-adjointness of the starting equation. The one- and two-term approximations are plotted in Figure 8.1 together with the exact solution. It is clear that for this simple problem the two-term approximation is sufficient.
   The application of Galerkin's method to a linear boundary value problem is particularly straightforward because the resulting equations for the coefficients

are linear and can easily be solved. Galerkin's method has therefore been used extensively for the solution of linear eigenvalue problems; the eigenvalue problem for a differential equation is transformed using Galerkin's method into a matrix eigenvalue problem, for which efficient computer programs are available. Most problems of interest are nonlinear, however. In that case the resulting algebraic equations for the coefficients are also nonlinear, and they may involve integrals that can be evaluated only numerically. The overall procedure is the same, but now the nonlinear equations for the coefficients must be solved iteratively, typically using a multidimensional Newton-Raphson approach. We first evaluate the residual using a nominal set of coefficients $\{\overline{C_n}\}$. The new estimates $C_n = \overline{C_n} + \delta C_n$ are then evaluated by linearizing about the nominal solution to obtain the linear equations

$$\sum_k A_{mk}\delta C_k = b_m, \tag{8.9a}$$

$$A_{mk} = \int \varphi_m \frac{\partial \Re}{\partial C_k} dx, \tag{8.9b}$$

$$b_m = -\int \varphi_m \Re dx, \tag{8.9c}$$

where the integrals are evaluated using the nominal values $\{\overline{C_n}\}$. The process is then repeated until the coefficients converge.

8.3 Finite Elements

As developed in the preceding section, Galerkin's method employs a single set of functions that approximate the solution over the entire spatial domain. This restriction is not necessary, and we could employ different functions in different parts of the domain. The formalism is a straightforward extension of what we have already done, but it is demanding in detail in a way that is inappropriate for our focus, so we will simply sketch out the idea without going into the specifics of implementation.

We divide the spatial domain into a set of subdomains (*finite elements*), and we define an approximate solution over each element; the approximation must usually be continuous across the element boundary. The elements need not all be the same size. By applying Galerkin's method we now obtain a set of algebraic equations for the coefficients associated with the approximate solution in each element, but the equations for the coefficients are coupled by the requirement that the solution be continuous in passing from one finite element to the next. This approach has an obvious advantage over the traditional application of Galerkin's method, in that the elements can be made large in regions where the function is changing slowly and small in regions where the function is changing rapidly. In that way, very simple approximating functions in each element – linear or quadratic functions in most cases – can be expected to provide good local approximations to the solution. The use of low-order approximations means that integrations can be carried out easily and in a general form that applies to any element.

Now, suppose we have a partial differential equation to solve in a two-dimensional domain. As in the one-dimensional problem, we divide the region into a set of M elements; quadrilaterals and triangles are the shapes most commonly used in two dimensions, and these are generally adequate for fitting complex shapes. We then use Galerkin's method to approximate the solution in each element, using a low-order polynomial as the approximating function. The sum of the number of equations resulting from setting weighted residuals to zero in all M elements, matching continuity conditions at all common faces, and satisfying essential boundary conditions equals the total number of unknown coefficients.

This sketchy description suggests the potential advantage of the finite element method over other approximate methods of the solution of partial differential equations. In the more familiar finite difference method, derivatives are approximated by differences of the values of the function between two different grid points, thus also generating a set of algebraic equations for the dependent variables at the spatial grid points. It is difficult to write a general finite difference code that permits different grid spacings in different regions in order to incorporate large distances between grid points where the function is changing slowly and small distances where the function is changing rapidly, however, whereas this is relatively straightforward with finite elements. Because of the universal nature of the integrals that arise when setting weighted residuals to zero, it is also straightforward to write general codes that permit efficient remeshing if the user wishes to re-solve the problem with a different distribution of elements. In addition, finite elements are very convenient for representing nonrectangular regions, including regions with odd shapes such as might be encountered in typical molding applications. Finite differences are more difficult to apply in such situations, although techniques do exist.

One typically uses quadratic functions in each element in two dimensions to approximate velocities and linear elements to approximate the pressure. There is a theoretical foundation for incompressible Newtonian fluid flows for this selection based on the fact that the incompressibility equation, which is associated with the pressure, acts as a constraint on the momentum equation in the variational formulation. Selecting the same order of approximation for the velocity and pressure would overconstrain the velocity field. The momentum equation for a Newtonian fluid involves a second derivative of the velocity with respect to the spatial coordinates, but a partial integration is performed in creating the weighted residual that leads to a first derivative; hence, only continuity of the approximating function is required, while the first-order derivative can be discontinuous across the element boundary. The partial integration also provides a natural way to impose force boundary conditions. It is obvious that the integrations to obtain the weighted averages of the residuals are straightforward with low-order polynomials, and they are the same for all interior elements of the same shape. The major problem in implementation is one of bookkeeping, since a rigorous system of numbering elements must be employed to maintain the generality that makes the method attractive.

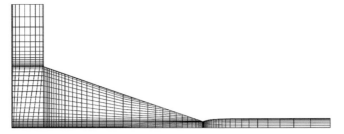

Figure 8.2. Finite element mesh for a sheet-coating die.

8.4 Sheet-Coating Die

We consider the sheet-coating die as a first example of the use of the finite element method to solve a processing problem. This process was analyzed in Section 5.3 using the lubrication approximation, where we found that there can be a region of backflow within the die. The finite element mesh is shown in Figure 8.2. In contrast to the analysis in Section 5.3, here we include an entry region orthogonal to the moving sheet, and we include the developing free surface of the film on the sheet after it has left the die. The free surface is calculated with an iterative scheme that includes the points defining the surface as unknowns and requires that the kinematical condition of no flow through the surface (compare Equation 7.17 in the spinline analysis) be satisfied; the final mesh is shown. The elements in the converging region of the die are smaller in the transverse direction close to the moving sheet than in the interior because the largest gradients in the velocity are expected below the region of recirculation. There is a singularity in the lower left-hand corner of the flow region, where the no-slip condition on the vertical wall meets the fixed velocity condition on the horizontal moving sheet; this is resolved in the numerical simulation shown here by having a small region near the corner where the wall velocity increases quadratically from zero to the velocity of the sheet.

The calculations are for an isothermal Newtonian fluid. Because of the quasi-linearity of the flow equations with respect to velocity, it is not necessary to specify absolute magnitudes for the dimensions and physical properties; all lengths and velocities simply scale linearly with the reference length and velocity, respectively, while the pressure and stresses scale linearly with a characteristic stress defined by the reference length and velocity and the magnitude of the viscosity. (It is necessary, of course, that the reference quantities make physical sense when converted to real-world dimensions; in particular, Reynolds numbers must be sufficiently low to ignore inertial effects, and shear rates must be sufficiently low to neglect viscous heating.) The values given here are dimensional, but other magnitudes can be accommodated by simply changing the scales of length, velocity, and viscosity. The geometry is as follows: exit gap spacing = 1 mm; width of the entry region = 5 mm; total height of the entry region above the moving sheet = 20 mm; height of the die where the upper wall intersects the entry region = 10 mm; length of the converging region of the die = 25 mm; length of the sheet after emerging from the die = 20 mm. The speed V of the moving plate is 1,000 mm/s, and the flow rate per unit width q is

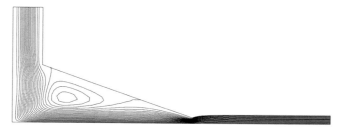

Figure 8.3. Streamlines for flow in a sheet-coating die.

1,500 mm^2/s; hence, the average velocity at the die exit is 1,500 mm/s, and the final film thickness is 1.5 mm. The viscosity is 30 Pa s; results of calculations with a viscosity equal to 30,000 Pa s were identical except for the magnitude of the pressure.

The computed streamlines are shown in Figure 8.3. The slight "wiggle" on one streamline is typical and reflects the discrete nature of the computation. The streamline that separates the recirculating region from the portion of the flow that exits the die is clearly seen; this corresponds to the line $H_o(x)$ in Figure 5.3. Pressure profiles are shown in Figure 8.4. The numerical solution incorporates features that are not accessible with the lubrication approximation. The latter is valid only as far back as the center of the recirculating region and cannot predict the closed streamlines or the region of rearrangement between the entering flow and the flow in the converging region of the die, nor can it predict the flow in the immediate neighborhood of the die exit. The lubrication approximation also predicts straight, vertical pressure contours, which are seen over only a portion of the flow regime. The pressure becomes infinite at the point where the melt separates from the die wall, but this is known to be an "integrable" singularity in which the force (the integrated pressure) over any region is finite; infinite stresses at a point are permitted in a continuum theory, but infinite forces, of course, are not. The pressure contours in the region of the die exit are shown on a magnified scale in Figure 8.5, where it is to be noted that large negative pressures (relative to atmospheric) will exist in a small region. Negative pressures of more than one atmosphere over a finite region in a quiescent fluid can cause cavitation, but the continuum equations used here do not address issues like nucleation and the growth of voids. (Care must be taken in any event in

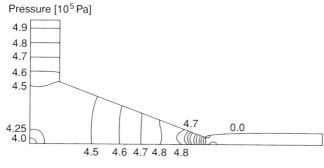

Figure 8.4. Computed pressure profiles in a sheet-coating die.

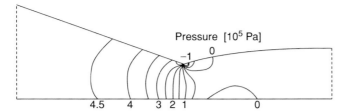

Figure 8.5. Pressure contours near the die exit.

interpreting negative pressure for a viscous incompressible fluid in motion, since the total stress is the sum of the isotropic pressure and the extra stress.)

8.5 Extruder

We now turn to the planar model of an extruder considered in Chapter 3. The dimensions and parameters are those used in Examples 3.1 and 3.2. A Newtonian melt with viscosity 30,000 Pa s and density 782 kg/m^3 is to be extruded through a slit die with a gap of 2.5 mm, a length of 75 mm, and a width of 1.5 m. The upstream plane channel is 0.75 m long and has a height of 20 mm. The geometry is shown drawn to scale in Figure 8.6. (Note the presence of a small vertical entry region at the upper left, which is not included in the analytical solution that assumes rectilinear flow. We will see that this detail is important.) The moving lower surface, which approximates the extruder screw, has a linear velocity of 12.4 mm/s. The melt is fed at a temperature of 140 °C. The upper surface is also maintained at 140 °C, as is the temperature along the vertical left-hand wall. (The former is equivalent to assuming an infinite Biot number, in which the wall temperature can be fixed directly, while the latter is intended to approximate a well-mixed feed stream with a uniform temperature.) The lower (moving) surface is adiabatic, as is the vertical surface at the entry to the die. The flow in the die is taken to be adiabatic. The pressure at the entrance and exit of the system is atmospheric; this is implemented in the computer code by setting the normal force to zero along these surfaces. The temperature gradient in the flow direction is set to zero at the exit of the die; this condition reflects the fact that there is no axial conduction and implicitly assumes that further vertical heat transfer beyond the exit will also be negligible. The die is sufficiently short that the specific exit boundary condition has little effect on the calculations. The no-slip condition is implemented for the velocity at all solid surfaces. We assume that the physical properties are independent of temperature, so the velocity field is uncoupled from the temperature field. A more realistic calculation would include at least the temperature dependence of the viscosity, but we wish to be able to make a direct comparison to the asymptotic solutions obtained for this system in Chapter 3.

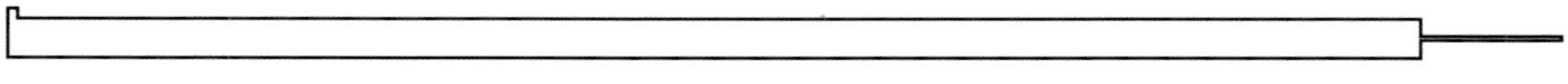

Figure 8.6. Geometry of the one-dimensional extruder, drawn to scale.

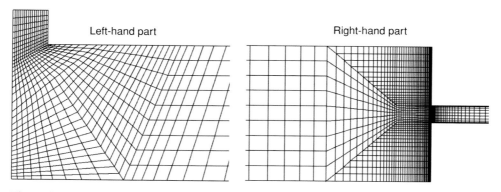

Figure 8.7. Finite element meshes for the left- and right-hand portions of a one-dimensional extruder.

The left- and right-hand portions of the finite element mesh are shown in Figure 8.7. The elements are smaller in regions where more rapid changes in the dependent variables are expected, with the densest mesh being used in the entry region to the die. The mesh shown here, using quadratic approximating functions for the velocities and temperature and linear interpolating functions for the pressure, leads to 44,389 coupled algebraic equations; a calculation using a mesh leading to 66,457 equations gave equivalent results. There is one important computational issue that was already mentioned in the previous example: There are different boundary conditions on the orthogonal surfaces arbitrarily close to the two lower corners of the upstream channel. In view of the quadratic interpolation selected for the velocity, the computational scheme cannot resolve the large gradients caused by the sudden transition from a zero velocity on the vertical surfaces to a finite velocity on the moving plate if the change is made only over the corner element, and spurious velocity oscillations result. Thus, a gradual transition was used for this calculation; for the examples shown here the velocity was changed quadratically from 0 to 12.4 over a length of 5 mm.

The computed streamlines are shown in Figure 8.8, with a magnified view of the exit region in Figure 8.9. (Recall that the relative horizontal and vertical scales in Figure 8.8 have been adjusted for clarity.) It is clear that the parallel flow assumed in the analysis in Chapter 3 exists over most of the region, with small regions near

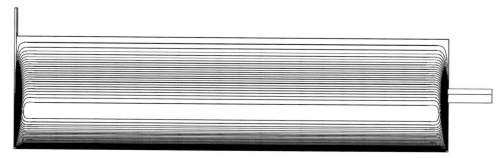

Figure 8.8. Computed streamlines for a planar extruder.

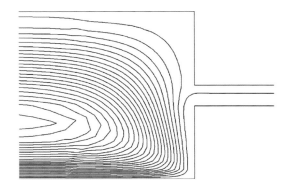

Figure 8.9. Streamlines in the exit region.

the ends of the upstream channel in which the flow reverses. Most of the fluid recirculates for the parameters used in this example, with the incoming melt moving rapidly downward to the moving plate, where it is then conveyed to the exit. Pressure contours, which are not shown, are vertical except in the immediate area of the exit. The pressure profile along the midplane is shown in Figure 8.10, and it is clear that the expected linear profile is obtained in both the upstream "extruder" channel and in the die. The computed maximum pressure is 3.97×10^6 Pa, which is slightly less than the value of 4.1×10^6 computed from the analytical solution. Similarly, the computed throughput is 2.26×10^{-6} m^2/s, compared to 2.37×10^{-6} from the analytical solution. These small differences are likely a consequence of the more complex flow in the finite geometry. (A calculation in which the melt simply enters from the left in parallel flow gives a flow rate of 2.32×10^{-6} m^2/s.) It is important to note one fundamental difference in the formulation between the analytical and numerical problems. In the analytical solution the flow rate is specified and the wall velocity is computed. In the numerical solution the wall velocity is specified and the flow rate is computed. The recirculating flow therefore has an effect on the flow rate in the numerical solution for a specified wall velocity.

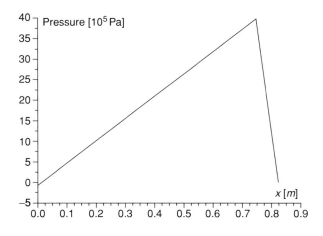

Figure 8.10. Pressure profile.

Figure 8.11. Computed temperature contours for a planar extruder. For a color version of this figure, please go to the plate section.

Temperature contours are shown in Figure 8.11, with a magnified image of the exit region and die in Figure 8.12. The temperature profile along the moving plate is shown in Figure 8.13a, and the profile along the midplane, including the die, is shown in Figure 8.13b. The effect of cooling along the upper wall is obvious, since the maximum temperature rise is only about 20 °C, in contrast to the computed adiabatic temperature rise of 140 °C for the analytical solution in Chapter 3. It is notable that the maximum temperature is in the interior, rather than at the adiabatic wall, over about the first two thirds of the channel; this is because the feed, which is at the lowest temperature in the system, is carried quickly to the moving plate. Thus, conclusions about the temperature distribution based on the assumption of a fully developed profile can be very misleading.

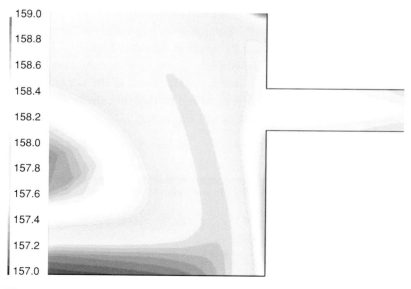

Figure 8.12. Computed temperature contours in the region near the die. For a color version of this figure, please go to the plate section.

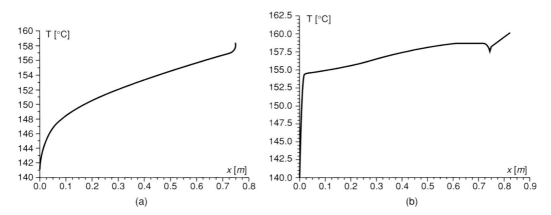

Figure 8.13. Computed temperature profile along the (a) moving plate and (b) center plane.

The assumption about the upstream conditions is crucial. If we assume instead that the fluid enters with parallel streamlines and that the axial temperature derivative at $x = 0$ is zero, we obtain the temperature map shown in Figure 8.14, which is very different qualitatively and quantitatively from the one in Figure 8.11. These boundary conditions are equivalent to assuming that the temperature profile is fully developed at the entrance (i.e., that the upstream channel is infinitely long), with the subsequent distortion in the final section caused by the recirculating flow at the exit. The computed temperature at the bottom plate at $x = 0$ is 196 °C, which is essentially the value obtained from Equation 3.34a with the parameters used here and in Chapter 3 if the heat transfer coefficient U is set to zero. (The limit corresponds to the case $Bi \to \infty$ and $\Pi \to 0$, with the product ΠBi remaining finite.) This computation emphasizes the importance of the proper problem formulation; the output from a computer simulation is only as meaningful as the input.

There is little difficulty in extending the computation to a non-Newtonian fluid. The recirculating region near the entrance to the die is shown in Figure 8.15 for the Newtonian fluid and for power-law fluids with $n = \frac{1}{2}$ and $\frac{1}{4}$. Shear thinning effects a qualitative change in the flow. The small recirculating region in the upper corner, which is sometimes known as a *Moffatt vortex* and can be theoretically predicted for corner flows, loses strength with decreasing power-law index.

Figure 8.14. Computed temperature profile with uniform flow from the left. For a color version of this figure, please go to the plate section.

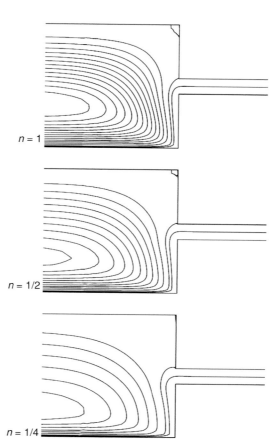

Figure 8.15. Flow near the die entry for power-law fluids.

8.6 Squeeze Flow and Fountain Flow

Squeeze flow between parallel plates was analyzed in Section 6.3 as an elementary model of compression molding. In that treatment we were able to obtain an analytical solution to the creeping flow equations for isothermal Newtonian fluids by making the kinematical assumption that the axial velocity is independent of radial position (or, equivalently, that material surfaces that are initially parallel to the plates remain parallel). In this section we show a finite element solution for non-isothermal squeeze flow of a Newtonian liquid. The geometry is shown schematically in Figure 8.16. We retain the inertial terms in the Navier-Stokes equations, thus including the velocity transient, and we solve the full transient equation for the temperature, including the viscous dissipation terms. The computational details,

Figure 8.16. Schematic of axisymmetric squeeze flow. Reprinted from Debbaut, *J. Non-Newtonian Fluid Mech.*, **98**, 15 (2001). Copyright Elsevier.

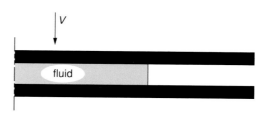

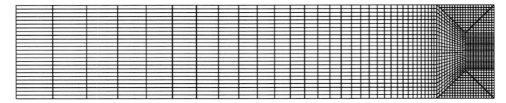

Figure 8.17. Finite element mesh for squeeze flow. Reprinted from Debbaut, *J. Non-Newtonian Fluid Mech.*, **98**, 15 (2001). Copyright Elsevier.

including the handling of the no-slip boundary condition and the transition to a stress-free boundary condition at the outer radial contact point, are described in the cited article by Debbaut. Solving the transient equations requires some modification of the basic Galerkin finite element method outlined in Section 8.2, and a special technique was used to permit the internal elements to deform as the flow progresses without leading to shapes that are inefficient for accurate computation. These details are beyond our consideration in this introductory treatment, however, and they are incorporated in any event into commercial finite element codes. The calculations were carried out using the mesh shown in Figure 8.17, which contains 1,746 elements.

The calculations shown here were carried out for a fluid with a constant viscosity of 1,000 Pa s, a density of 1,000 kg/m^3, and a heat capacity and thermal conductivity of 2,000 J/kg/K and 0.5 W/m/K, respectively. All calculations were started from rest at a uniform temperature of 200 °C, a plate spacing of 1 cm, and a sample radius of 5 cm, with the upper plate moving at a constant downward speed of 10 cm/s and the lower plate stationary. Figure 8.18 shows the sample shape and temperature contours as functions of time for a calculation in which the plates are assumed to

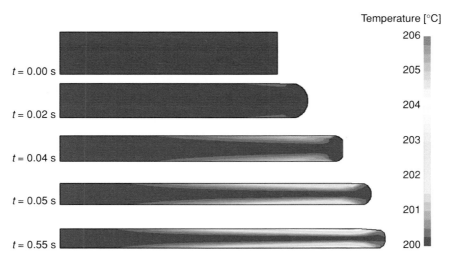

Figure 8.18. Shape and temperature evolution in time for adiabatic plates. Reprinted from Debbaut, *J. Non-Newtonian Fluid Mech.*, **98**, 15 (2001). Copyright Elsevier. For a color version of this figure, please go to the plate section.

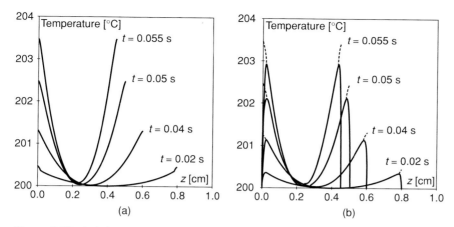

Figure 8.19. Axial temperature profiles 5 cm from the center: (a) adiabatic plates; (b) plates maintained at 200 °C (dashed lines are adiabatic curves). Reprinted from Debbaut, *J. Non-Newtonian Fluid Mech.*, **98**, 15 (2001). Copyright Elsevier.

be adiabatic. Shear-induced temperature boundary layers develop rapidly near the outer edge. (There is a negligible temperature rise for these parameters if the no-slip boundary condition is replaced by perfect slip.) Axial temperature profiles are shown in Figure 8.19a at a distance of 5 cm from the center. Figure 8.19b shows the development of the temperature profiles for a case in which the fluid temperature at the plates was fixed at 200 °C; the dashed lines in Figure 8.19b are the adiabatic curves, which are nearly the same as those with fixed temperature (infinite heat transfer coefficient) except in the thin layer immediately adjacent to the wall. The calculations were not carried out for times longer than 0.055 s because of the distortion of the mesh. It should be recalled that the flow equations are decoupled from the temperature profile development because of the assumption of a constant viscosity, but the relatively small localized temperature increases from dissipation for the assumed viscosity will not have a major effect on the flow.

The basic assumption in the analytical solution to the creeping flow equations is that material planes that are initially parallel to the plates remain parallel. The deformation of material planes is shown in Figure 8.20. (Although the calculation includes the inertial terms, the effect of inertia is small after an initial transient, so the comparison is meaningful.) The material planes do remain parallel over much of the original sample radius, but there is substantial deformation near the outer edge, where the *fountain flow* caused by the no-slip condition can be seen. Fountain flow is the name given to the stagnation flow near the free surface. Because of the no-slip condition, fluid cannot move along the wall to fill the mold; hence, fluid must move to the wall from the bulk. The effect can also be seen in the contours of the axial velocity at $t = 0.02$ s shown in Figure 8.21, where the axial velocity is independent of radial position except near the free surface. Fountain flow exists in all mold-filling operations and is a major factor in the development of morphology. The flow field near the stagnation point is purely extensional, for example; extensional flow is much more efficient at aligning polymer chains and glass or carbon fibers than shear

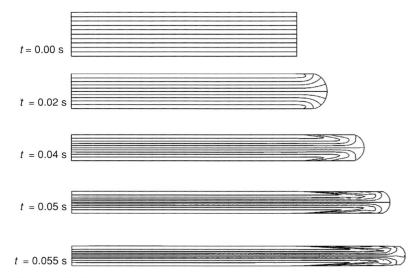

Figure 8.20. Deformation of material planes that were originally horizontal. Reprinted from Debbaut, *J. Non-Newtonian Fluid Mech.*, **98**, 15 (2001). Copyright Elsevier.

flow, so the fluid deposited on the plate that originated in the interfacial region is likely to be much more oriented than the material in the core. The fountain flow is therefore the cause of layered morphologies that are often observed in molded parts.

8.7 Concluding Remarks

Computational fluid dynamics now plays a major role in process and product design in many fields, including, of course, polymer processing. The basic concepts are straightforward, but the details are complex and require the study of specialized texts and the periodical literature for implementation. Commercial codes are designed with user-friendly "front ends" that facilitate use without a detailed understanding of the methodology; this simplicity is a mixed blessing, since issues that are related to the computational methodology rather than the underlying physics may be masked, and spurious results can sometimes be obtained.

The examples in this chapter were selected to illustrate the way in which numerical simulation complements and enhances the understanding of polymer processing operations that is gained from analytical solutions based on idealizations (infinite geometries and the lubrication approximation, for example). The analytical solutions are invaluable for providing insight, but detailed information and complete

V_z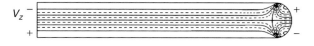

Figure 8.21. Contours of axial velocity at $t = 0.02$ s. The interval between contours is 2 cm/s. Reprinted from Debbaut, *J. Non-Newtonian Fluid Mech.*, **98**, 15 (2001). Copyright Elsevier.

understanding in all but the simplest geometries can come only from the numerics; this is especially true when there are major thermal effects.

Until this point we have considered only inelastic liquids in both the analytical and numerical treatments of polymer processing. The viscoelasticity of polymer melts sometimes plays a major role in the mechanics of processing behavior, and we take up this important issue in the next and subsequent chapters. Numerical problems are greatly compounded by the presence of fluid elasticity, but the overall approach is unchanged. We will return to the use of computational fluid dynamics with complex rheology after taking up the subject of viscoelasticity.

BIBLIOGRAPHICAL NOTES

The use of Galerkin's method and the relation to the calculus of variations is treated, for example, in

Denn, M. M., *Stability of Reaction and Transport Processes*, Prentice Hall, Englewood Cliffs, NJ, 1975.
Finlayson, B. A., *The Method of Weighted Residuals and Variational Principles*, Academic Press, San Diego, CA, 1972.

The detailed application of the finite element method is covered in texts such as

Bathe, K.-J., *Finite Element Procedures,* 2nd ed., Prentice Hall, Upper Saddle River, NJ, 1995.
Reddy, J. N., *An Introduction to the Finite Element Method*, McGraw-Hill, New York, 1984.
Reddy, J. N., and D. K. Gartling, *The Finite Element Method in Heat Transfer and Fluid Mechanics*, CRC Press, Boca Raton, FL, 1994.
Zienkiewicz, O. C., *The Finite Element Method*, 3rd ed., McGraw-Hill, New York, 1977.

The squeeze-flow application in Section 8.6 is from

Debbaut, B., *J. Non-Newtonian Fluid Mech.*, **98**, 15 (2001).

A finite-element solution for squeezing flow of two layered Newtonian fluids with different viscosities, together with a lubrication-type analytical solution, is in

Lee, S. J., M. M. Denn, M. J. Crochet, and A. B. Metzner, *J. Non-Newtonian Fluid Mech.*, **10**, 3 (2001).

Fiber spinning of a Newtonian liquid (Chapter 7), including the spinneret flow and the region of velocity rearrangement, is studied using finite elements in

Fisher, R. J., M. M. Denn, and R. I. Tanner, *Ind. Eng. Chem. Fundam.* **19**, 195 (1980).

Hollow fiber spinning of a Newtonian liquid, in which there are two free surfaces, is discussed in

Freeman, B. D., M. M. Denn, R. Keunings, G. E. Molau, and J. Ramos, *J. Polym. Eng.*, **6**, 171 (1986).
Su, Y., G. G. Lipscomb, H. Balasubramanian, and D. R. Lloyd, *AIChE J.*, **52**, 2072 (2006).

Other applications to fluid mechanics and polymer processing problems can be found on a regular basis in publications such as the *Journal of Fluid Mechanics*, the *Journal of Non-Newtonian Fluid Mechanics*, *International Polymer Processing*, and so forth. We address polymer processing flows in which polymer viscoelasticity is important in Chapter 10.

9 Polymer Melt Rheology

9.1 Introduction

Our analysis of polymer melt processing operations has thus far assumed that the polymer melt can be described as an *inelastic* liquid, and in fact we have generally assumed for simplicity that the melt is Newtonian. An inelastic liquid has no memory; that is, the stress in the fluid at a given time and place depends only on the deformation rate at that time and place. Entangled polymers *should* have memory, since the response to a deformation must depend on the reorganization of the entangled macromolecules, which cannot be instantaneous. We saw a manifestation of such memory in Figure 1.8, where a silicone polymer being squeezed between two plates under constant force "bounced," causing transient *increases* in the gap spacing. Another way to think about memory is to imagine the polymer melt at rest, with the chains forming an entangled network. The chains cannot respond instantaneously if we attempt to deform the melt rapidly because they are entangled, so the initial short-time response must be that of a rubberlike network, not a viscous fluid, including shape recovery if the stress causing the deformation is quickly removed. In general, we expect to see a superposition of two responses: the short-time rubberlike response caused by deformation of the entangled network and the long-time viscous response caused by the dissipative process of relative chain motion in the flowing melt. Hence, polymer melts are *viscoelastic* liquids.

Rheology is the study of the response of materials to deformation, and one of the general goals of the discipline is to obtain the appropriate stress-deformation *constitutive equation* for a given material. Polymer melt rheology is a subject with an enormous literature that has developed over five decades. Constitutive equations for melts range from the phenomenological to the molecular. Some are available as differential or integral equations that relate the stress state directly to the deformation rate, hence permitting (in principle) simultaneous solution with the momentum, continuity, and energy equations. Others are indirect, sometimes requiring the solution of subsidiary equations describing the chain morphology, perhaps in the form of an equation for the distribution function of chain orientations. Modeling applications in polymer processing have been restricted to continuum equations that can be solved together with the conservation equations, and many applications have been further restricted to constitutive equations that can be expressed as differential equations. The preference for the differential equation form over the integral equation is mainly associated with technical details of the computation.

Our goal in this chapter is modest. We will consider the most common experimental manifestations of viscoelasticity in shear and elongational flows. We will then examine the most elementary forms for constitutive equations that can represent this behavior, namely, a class of equations consisting of generalizations of the *Maxwell model*. The generalized Maxwell models do a reasonable job of describing the behavior of polymer melts, and they have been used extensively in simulations. They contain little or no molecular structure information, however, unlike more recent continuum theories that start from a more complete molecular framework, but the transition to the use of molecularly based continuum theories in simulations is largely a matter of detail at the programming level.

9.2 Linear Viscoelasticity

9.2.1 Linear Dynamical Systems

We focus initially on small deformations of the entangled melt, which is the province of *linear viscoelasticity*. While processing operations clearly involve large deformations, we can learn a great deal about the melt response, even to large deformations, by probing the linear regime. The framework of linear viscoelasticity is identical to the frequency response methodology used in many other fields, including process dynamics and control.

Consider any dynamical system that can be characterized by an input $a(t)$ and an output $b(t)$, where a and b are functions of the independent variable, which we will assume is time. The input is often called the *forcing function*, and the output the *response*. The system is linear if the following criterion is satisfied:

1. input $\quad a_1(t) \Rightarrow$ output $b_1(t)$;
2. input $\quad a_2(t) \Rightarrow$ output $b_2(t)$.

   Then
3. input $\quad a_1(t) + a_2(t) \Rightarrow$ output $b_1(t) + b_2(t)$.

   It further follows from this relation that

4. input $\quad \alpha_1 a_1(t) + \alpha_2 a_2(t) + \cdots + \alpha_N a_N(t) \Rightarrow$ output $\alpha_1 b_1(t) + \alpha_2 b_2(t) + \cdots + \alpha_N b_N(t)$.

Integration and differentiation are linear operations, so processes described by linear differential and integral equations are linear dynamical systems.

An *autonomous* linear system is one in which all system parameters are independent of time, and integrals that exist in the system description are of convolution type (a convolution integral has the form $\int_0^\infty f(t - t') a(t') dt'$). It is readily

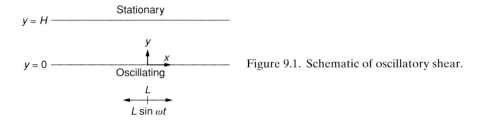

Figure 9.1. Schematic of oscillatory shear.

established that an input $a(t) = A \sin \omega t$ to any autonomous linear dynamical system results in an output $b(t) = B \sin (\omega t + \delta)$. The ratio B/A is known as the *amplitude ratio*, which we will denote AR. The amplitude ratio AR and the phase angle δ are functions of the forcing frequency, ω. There are several equivalent ways of representing the output $b(t)$:

$$b(t) = B(\omega)\sin(\omega t + \delta) = B(\omega)\cos\delta\sin\omega t + B(\omega)\sin\delta\cos\omega t$$
$$= C(\omega)\sin\omega t + D(\omega)\cos\omega t, \tag{9.1}$$

$$AR = B/A = \left(C^2 + D^2\right)^{1/2}/A, \tag{9.2a}$$

$$\tan\delta = D/C. \tag{9.2b}$$

It can be shown that knowledge of $AR(\omega)$ and $\delta(\omega)$, or, equivalently, of $C(\omega)$ and $D(\omega)$, for all frequencies ω is equivalent to complete knowledge of the equations describing the dynamical system. (Readers who have studied Fourier analysis will recognize that this is simply a statement that the same information is contained in a function and its Fourier transform.) In the discipline of process dynamics and control, it is common to represent system behavior with the functions $AR(\omega)$ and $\delta(\omega)$, usually in the form of a *Bode diagram*, in which the logarithm of AR is plotted versus the logarithm of ω and δ is plotted on a linear scale versus the logarithm of ω. Data in the field of process control are sometimes represented on a *Nyquist diagram*, in which C is plotted on a linear scale versus D, with ω as a parameter; this representation is also known as a *Cole-Cole plot*, and the latter name is commonly used in the spectroscopy and rheology literature.

Physical systems are typically nonlinear. For sufficiently small variations about an equilibrium state, however, most systems will respond in a linear manner, with an error that is on the order of the amplitude of the input. The experimental test of linearity is to ensure that the amplitude ratio remains constant with changes in the amplitude of the input.

9.2.2 Linear Viscoelastic Functions

Now let us consider the mechanical experiment shown schematically in Figure 9.1. A material, fluid or solid, is placed between two plates that are separated by a distance H. One plate is oscillated sinusoidally relative to the other, with amplitude L and frequency ω. We measure the shear stress τ on the upper plate as a function of time. (The experiment is usually done by rotating one of the plates around an axis, rather than by imposing a linear motion, so the stress is determined from a torque measurement. In fact, the rotational instrument often employs a cone and a plate or concentric cylinders, rather than parallel plates. These are details that do not concern us, however, since it is readily shown that the same physical quantity is measured in all these geometries.) The *shear strain* γ is the displacement normalized with respect to the gap spacing. The input $a(t)$ is therefore $\gamma(t) = (L/H) \sin \omega t$, while the output $b(t)$ is the shear stress, $\tau(t)$. It is useful to note that the rate of shear strain, which is called both the *shear rate* and the *strain rate* interchangeably, is $(L\omega/H) \cos \omega t$.

We test the experimental system to ensure that the imposed strains are in the linear regime. The output will then be in the form of Equation 9.1 as follows:

$$\tau(t) = G' [(L/H) \sin \omega t] + G'' [(L/H) \cos \omega t]. \tag{9.3}$$

The "prime" notation G' and G'' is conventional and does *not* mean differentiation. $G'(\omega)$ is known as the *storage modulus*, and $G''(\omega)$ is known as the *loss modulus*. The nomenclature follows from our understanding of classical materials. A linear elastic material (a *Hookean solid*) is a material for which the stress is proportional to the strain, and the deformation is completely recoverable; that is, the energy required for displacement is stored elastically, and the body returns to its undeformed shape when the stress is removed. The stress for a linear elastic material will be *in phase* with the strain in an oscillatory experiment. Hence, G' defines the magnitude of an elastic response to a deformation. A linear viscous material (a *Newtonian liquid*) is a material for which the stress is proportional to the strain rate, and the deformation is completely nonrecoverable; that is, the energy required for displacement is dissipated, and the body remains deformed when the stress is removed. The stress for a linear viscous material will be in phase with the strain rate, or 90° *out of phase* with the strain.

Alternatively, the output is sometimes written

$$\tau(t) = \eta'' [(L\omega/H) \sin \omega t] + \eta' [(L\omega/H) \cos \omega t]. \tag{9.4}$$

$\eta'(\omega)$ is known as the *dynamic viscosity*; it is the coefficient of the term proportional to the strain rate; hence, it is a measure of the dissipative response. $\eta''(\omega)$ is rarely used. Clearly, $\eta' = G''/\omega$, while $\eta'' = G'/\omega$. Still another equivalent form for the output is

$$\tau(t) = G' [(L/H) \sin \omega t] + \eta' [(L\omega/H) \cos \omega t] = G'\gamma + \eta' d\gamma/dt. \tag{9.5}$$

Finally, it follows that

$$\tan \delta = G''/G' = \eta'/\eta''. \tag{9.6}$$

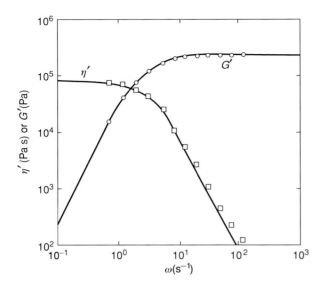

Figure 9.2. Linear viscoelastic response of a silicone polymer. Data of F. N. Cogswell.

The amplitude ratio is equal to $\omega|\eta^*|$, where the *magnitude of the complex viscosity* $|\eta^*|$ is defined

$$|\eta^*| = [(\eta')^2 + (\eta'')^2]^{1/2}. \tag{9.7a}$$

Similarly,

$$|G^*| = [(G')^2 + (G'')^2]^{1/2}. \tag{9.7b}$$

Only two of the many functions defined here are independent, but all are used. The input is sometimes taken to be a cosine rather than a sine, in which case there may be differences in algebraic signs in some of the equations, but this is not important. The functions G', G'', η', and η'' are always positive. (Positivity of G'' and η' is required by the second law of thermodynamics. There does not seem to be a thermodynamic requirement for positivity of G'.)

   Figure 9.2 shows experimental data for a silicone polymer similar to the one used in the squeeze flow experiment shown in Figure 1.9. The material is *viscoelastic*, since both the storage modulus and the dynamic viscosity are nonzero. At low frequencies the storage modulus goes to zero and the dynamic viscosity goes to a low-frequency asymptotic value. The deformation at low frequencies is sufficiently slow to allow the individual polymer chains to respond to the imposed strain; hence, the response is viscous, and it is clear that the low frequency limit of η' must be the zero-shear viscosity, η_0. At high frequencies the individual chains are unable to respond and the stress is entirely the consequence of deformation of the entangled network. In this limit the polymer melt is indistinguishable from a cross-linked rubber network, and the deformation is that of an elastic body, with G' going to an asymptotic value and η' to zero. The value of G' in this *rubbery plateau* region is known as the shear modulus and is usually denoted G.

   The data in Figure 9.2 are atypical in one important regard. The transition between the low-frequency viscous response and the high-frequency elastic

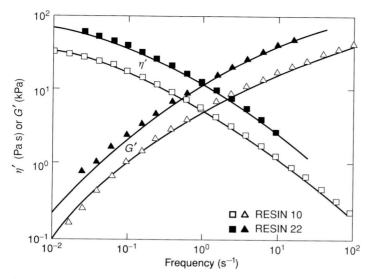

Figure 9.3. Linear viscoelastic response of two polyethylene resins at 170 °C. Reprinted with permission from Tsang and Dealy, *J. Non-Newtonian Fluid Mech.*, **9**, 203 (1981). Copyright Elsevier.

response is very sharp, occurring over about one decade of frequency. The data in Figure 9.3, for two polyethylene (PE) resins studied by Tsang and Dealy, with properties listed in Table 9.1, are more typical of commercial polymers. The transition region is very broad, reflecting the broad molecular weight distribution of the melts, and it is sometimes difficult to reach either the low frequency or rubbery plateau regions with commercial instrumentation.

9.2.3 Relaxation Spectrum

The data in Figure 9.2 are characteristic of a first-order process, in which the dynamical response is described by a single first-order linear differential equation. If we write the equation for the stress as

$$\lambda \frac{d\tau}{dt} + \tau = \eta \frac{d\gamma}{dt}, \tag{9.8}$$

Table 9.1. *Properties of polyethylene resins*

McGill stock number	10	22
Manufacturer	Union Carbide Canada	DuPont of Canada
Trade name/number	DFDQ4400	Sclair 59C
Type	Film resin (highly branched)	Blow molding resin (linear)
Solid density (kg/m^3)	915 (low-density PE)	960 (high-density PE)

Data of Tsang and Dealy, *J. Non-Newtonian Fluid Mech.*, **9**, 203 (1981).

then it readily follows that

$$\eta' = \frac{\eta}{1 + \lambda^2 \omega^2}, \qquad G' = \frac{\eta}{\lambda} \frac{\lambda^2 \omega^2}{1 + \lambda^2 \omega^2}. \qquad (9.9\text{a,b})$$

The solid lines through the data in Figure 9.2 are computed from Equations 9.9a–b with $\eta = 8 \times 10^4$ Pa s and $\lambda = 0.31$ s. (The silicone polymer in Figure 1.9 had $\eta = 6.65 \times 10^4$ Pa s and $\lambda = 0.1$ s.) The ratio η/λ is dimensionally a stress, corresponding to the shear modulus, G. Equation 9.8 is known as a *linear Maxwell model*, after the nineteenth-century Scottish physicist James Clerk Maxwell, who first proposed such an equation to describe the dynamics of gases. We recover the one-dimensional description of a Newtonian fluid, in which the shear stress is proportional to the shear rate, for $d\tau/dt \ll \tau/\lambda$. We recover the one-dimensional description of a Hookean solid in differentiated form, in which the rate of change of the stress is proportional to the rate of change of the strain, for $d\tau/dt \gg \eta/\lambda$. The Maxwell model is a linear superposition of the two limits.

It is clear that Equation 9.8 cannot describe the data in Figure 9.3. It is customary to view the dynamics of the polymer melt as being made up of a number of modes, each of which is described by an equation of the form of Equation 9.8. We thus write the total shear stress τ as a sum of partial stresses,

$$\tau = \sum_n \tau_n, \qquad (9.10\text{a})$$

where each τ_n satisfies an equation of the form

$$\lambda_n \, d\tau_n/dt + \tau_n = \lambda_n G_n d\gamma/dt. \qquad (9.10\text{b})$$

The functions η' and G' are then easily shown to satisfy

$$\eta' = \sum_n \frac{\lambda_n G_n}{1 + \lambda_n^2 \omega^2}, \qquad (9.11\text{a})$$

$$G' = \sum_n \frac{G_n \lambda_n^2 \omega^2}{1 + \lambda_n^2 \omega^2}. \qquad (9.11\text{b})$$

(It is customary to use λ_n and G_n as the parameters in the equation, rather than λ_n and η_n; $\eta_n = \lambda_n G_n$.) This representation is simply a finite Fourier transform if the λ_n are calculated from the zeros of the trigonometric functions. The multiple modes follow from the different motions of segments of various lengths on individual chains, as well as the distribution of chain lengths in the melt, and molecular theories of polymer dynamics predict specific relations between the coefficients for different modes. It suffices for our purposes to think of the set $\{\lambda_n, G_n\}$ as fitting parameters. What is usually done is to select the time constants, typically with an even distribution on a logarithmic scale, and to fit the moduli. The allowable range of the time constants is clearly limited to the interval bounded by the reciprocals of the highest and lowest frequencies for which data are available, since this is the only range in which the dynamics have been probed. (The limits can be made more precise, and are actually somewhat narrower, but that is a technical detail that goes

Table 9.2. *Discrete spectra for data of*
Tsang and Dealy in Figure 9.3

λ_n (s)	G_n (Pa)	
	Resin 10	Resin 22
0.00316	20,954	
0.01	15,895	31,715
0.0316	11,783	26,379
0.1	8,342	19,782
0.316	5,386	13,221
1.0	3,245	7,966
3.16	1,703	4,086
10.0	797.6	1,703
31.6	317.2	590.6
100.0	8.54	166.5

beyond the issues we are considering here.) Fits to the data in Figure 9.3 are shown in Table 9.2. Most fits employed in practice use fewer modes. The solid lines in Figure 9.3 were computed from these parameters.

The storage and loss moduli are not independent functions. We define the *relaxation modulus* $G(t)$ through an experiment in which a sample is subjected to an instantaneous infinitesimal strain γ, after which the stress is allowed to relax. $G(t)$ is the ratio of the time-dependent stress to the strain γ. It can readily be shown that the two moduli are the Fourier sine and cosine transforms of $G(t)$:

$$G' = \omega \int_0^\infty G(t) \sin \omega t \, dt, \qquad (9.12a)$$

$$G'' = \omega \int_0^\infty G(t) \cos \omega t \, dt. \qquad (9.12b)$$

Equations 9.11a–b represent the case in which the relaxation modulus $G(t)$ is fit with a series of exponentials. $G(t)$ can be obtained in principle from the inverse transform of either G' or G'', but the inversion is an ill-defined problem because of the finite frequency range for which data are available.

It is often convenient to use complex number notation and define the complex modulus $G^*(\omega) = G'(\omega) + iG''(\omega)$, where $i^2 = -1$. Equations 9.12a–b can then be written concisely as

$$G^*(\omega) = G'(\omega) + iG''(\omega) = i\omega \int_0^\infty G(t)e^{-i\omega t} \, dt, \qquad (9.12c)$$

where we have made use of the fact that $e^{i\omega t} = \cos \omega t + i \sin \omega t$. There are related linear viscoelastic functions derived from other experiments, but we will not pursue them here.

9.2.3 Memory Integral Formulation

It is straightforward to show by integration that Equations 9.10a–b are equivalent
to the integral equation

$$\tau(t) = \int_{-\infty}^{t} \sum_{n} \frac{G_n}{\lambda_n} e^{-(t-t')/\lambda_n} \gamma(t')\, dt'. \tag{9.13}$$

Here the strain $\gamma(t')$ is taken relative to the current state, so $\gamma(t) = 0$. This *memory
integral* formulation is a special case of *Boltzmann's superposition integral*,

$$\tau(t) = \int_{-\infty}^{t} m(t-t')\gamma(t')\, dt'. \tag{9.14}$$

The equivalence between the differential and integral equation formulations exists
only when the memory function $m(t)$ is expressed as a sum of one or more exponentials, but the exponential representation is invariably used for the memory function.

9.2.4 Time–Temperature Superposition

The dynamics of polymer chains are strong functions of temperature; chains move
easily at temperatures well above the glass transition temperature, while motion
becomes increasingly more difficult as the glass transition is approached. The effect
of temperature is to change the free volume in the melt, hence increasing or decreasing the mobility of the polymer chains. This obvious observation can be quantified
by use of the *time–temperature superposition* principle, which states that the *only*
effect of temperature on the melt is to scale the time axis. Thus, if we take linear
viscoelastic data as functions of frequency at a variety of temperatures, we expect
the data to superimpose onto master curves with the correct time scaling. The scaling function is usually denoted $a_T(T)$ and is called the *shift factor*. Since the low-
frequency asymptote of G'' equals the product of ω with the zero-shear viscosity
η_o, it follows that the shift factor has the temperature dependence of the zero-shear
viscosity.

Figures 9.4 and 9.5 show G' and G'' for two linear polymers, a poly(vinyl methyl
ether) (PVME) with a molecular weight of 138,000 and a polystyrene (PS) with a
molecular weight of 123,000, respectively. The data for each polymer have been
moved horizontally along the frequency axis until they form a single curve. There
is a substantial region of overlap, extending over three decades of frequency, so the
superposition is clearly established. The shift factors needed to obtain overlap of the
curves are shown as inserts. The reference temperature for each case was taken to
be 84 °C; this temperature has no significance other than being a convenient value
for the particular application for which the data were obtained, which was a study
of phase separation in blends of the two polymers. One of the significant uses of
time–temperature superposition is made evident by focusing on the open and closed
symbols in the PVME curves. The dynamic moduli are available over five orders

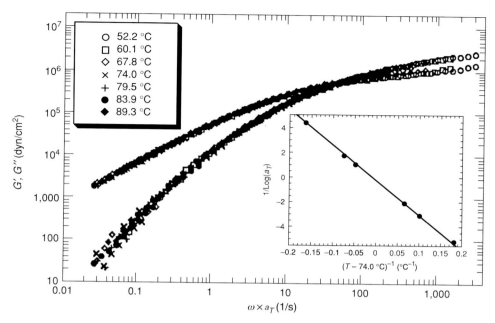

Figure 9.4. Master curves for a poly(vinyl methyl ether) with a molecular weight of 138,000. Data of Diana Hess.

of magnitude, while the range for any single temperature is considerably smaller. Even in this case the rubbery plateau has not been reached for PVME, although the low-frequency asymptotes ($G' \sim \omega^2$, $G'' \sim \omega$) have been achieved. The polystyrene data illustrate another important point. The reduced curve for G'' goes through a maximum and starts to decrease, as required by Equation 9.11a. (Recall that $G'' = \omega\eta'$.) There is then an upturn in the curve that is not predicted by the equation, and a second crossover between G' and G''; these frequencies are probing very short time scales beyond the rubbery region and reflect the start of a glassy response by the chains.

There are two common ways of representing the temperature dependence of the shift factor. One is as an exponential in reciprocal absolute temperature [$\exp(E/RT)$], as done with the polystyrene data; this functionality is usually valid at temperatures well above the glass transition temperature, which is 104 °C for the polystyrene. The polystyrene data are fit with $E/R = 9{,}300$ K; a value of order 4,000 K is typical of polyethylene, reflecting the wide variation in this parameter. The other commonly used functionality is the *Williams-Landel-Ferry* (WLF) equation, which is written

$$\log a_T = \frac{-C_1^o (T - T_o)}{C_2^o + T - T_o}. \tag{9.15}$$

T_o is the reference temperature. The WLF equation follows from free volume considerations and is usually used within about 100 °C of the glass transition temperature. The data for the PVME are seen to follow the WLF equation quite well, with $C_1^o = 7.3$ and $C_2^o = 220$ K. (There are "universal" values of the WLF parameters

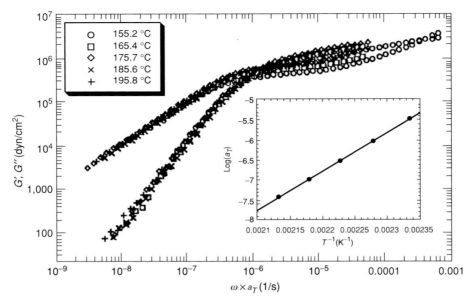

Figure 9.5. Master curves for a polystyrene with a molecular weight of 123,000. Data of Diana Hess.

that can be used when no data are available.) The free volume theory also accounts for the effect of pressure; hence, one expects to find a pressure shift factor a_P as well, and it does indeed exist. This is the source of the term $\exp(\beta p)$ in the pressure-dependent viscosity in Equation 4.1.

The data are sometimes shifted vertically by a ratio $\rho_0 T_0/\rho T$, where ρ is the density. This temperature dependence follows from the theory of rubber elasticity, in which the modulus is proportional to absolute temperature. The vertical shift is generally small and is rarely applied.

9.3 Shear Rheology

9.3.1 Normal Stress Difference

Polymer melts undergo large deformations in steady shear, and nonlinear effects associated with the dynamics of the polymer chains are expected. The existence of a finite *normal stress* is one manifestation of such nonlinearity. Suppose we shear a liquid between two parallel plates, one of which is moved relative to the other. The shear stress is equal to the ratio of the force required to move the plate at a given rate to the plate area, and it is from this measurement that we obtain the viscosity. With a polymeric liquid we find that a finite force is required to keep the spacing between the plates constant; in the absence of such a force the plates will move apart. A stress normal to the direction of shear is not observed for inelastic liquids. (A similar phenomenon is well known in the mechanics of solids. If a rod of

Figure 9.6. Schematic of rectilinear shear flow.

a rubbery material is twisted, there is a stress in the axial direction, orthogonal to the plane of deformation. This phenomenon is known as the *Poynting effect*.)

Consider the geometry shown in Figure 9.6. The velocity is in the "1" direction and the velocity gradient in the "2" direction; "3" is a neutral direction, sometimes known as the *vorticity* direction because it is the direction of the vorticity vector. The stress acting to push the plates apart in shear is $\sigma_{22} = -p + \tau_{22}$; the stress required to keep the plates together is $-\sigma_{22}$. We assume the velocity is only in the 1 direction, with a gradient only in the 2 direction, in which case it follows immediately from the momentum equation that σ_{11} and σ_{22} are constant throughout the flow field. The pressure in an incompressible liquid is a dynamical variable that needs to be eliminated in order to relate the stress to the deformation rate. To do this, we carry out an analysis analogous to the force calculation for the squeeze flow in Section 6.3. The stress at the edge, where the melt meets the atmosphere, must balance atmospheric pressure, which we can take to be zero. (We neglect the surface tension contribution, which will generally be negligible for polymer melts.) Thus, $\sigma_{11} = -p + \tau_{11}$, the stress in the flow direction, must equal zero at the interface, in which case it must be zero everywhere. Hence, $p = \tau_{11}$ and the stress required to keep the plates from separating is $\tau_{11} - \tau_{22}$. This normal stress difference is usually denoted N_1 and is called the *first normal stress difference*. N_1 is a function of shear rate; it can be shown from general continuum principles that N_1 is a quadratic function of shear rate at low shear rates and that N_1 plotted versus the square of the shear rate at small shear rates must be identical to $2G'$ plotted versus the square of the frequency at low frequencies. A *second normal stress difference*, $N_2 = \tau_{22} - \tau_{33}$, is usually numerically small relative to N_1 and of the opposite algebraic sign.

9.3.2 Measurement

Any flow field in which an orthogonal coordinate system can be embedded at each point along a streamline such that the velocity is a function of only one orthogonal coordinate direction and the magnitudes of the velocity and the velocity gradient are constant along the streamline can be shown to be mathematically equivalent to the rectilinear flow in Figure 9.6, and symmetry arguments can be used to show that there are three independent functions of the second invariant of the rate of deformation that are the same in all such flows: the viscosity, η, and the two normal stress differences, N_1 and N_2. Such geometries include cylindrical and plane channels in which drag or a pressure difference is employed to effect the flow, and a variety of torsional flows, including parallel plates, cone and plate, and concentric cylinders. Such flow fields are known as *viscometric flows* because they comprise all the flows

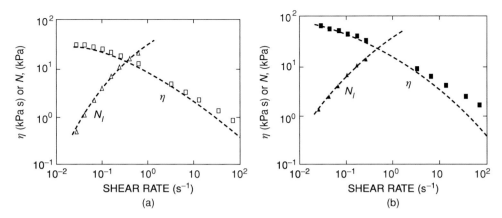

Figure 9.7. Steady shear and normal stress data at 170 °C for two polyethylenes: (a) low-density Resin 10 and (b) high-density Resin 22. The lines are the fit to the Phan-Thien/Tanner model, discussed below. Reprinted with permission from Tsang and Dealy, *J. Non-Newtonian Fluid Mech.,* **9**, 203 (1981). Copyright Elsevier.

used in viscometry; they are also known in the continuum mechanics literature as *flows with constant stretch history*, a categorization that includes a few flows in addition to the ones cited here.

There are important experimental differences between the various viscometric flows, despite their mathematical equivalence. The torsional geometries require only a few grams of polymer, while the pressure-driven flows require tens of grams. Torsional flows of polymer melts tend to become unstable at shear rates on the order of $1 \, \mathrm{s}^{-1}$, however, which is orders of magnitude below the range of interest for most processing applications. N_1 and N_2 can be measured in torsional flows, although the measurement of N_2 is quite difficult because of the small magnitude. The pressure drop in a capillary or slit is typically used to measure the viscosity at high rates; there is no established way to measure normal stresses in a capillary or slit, although there have been attempts based on approximate theories, and instrumentation is available. The major experimental problem with capillary and slit flow for measuring the viscosity of melts has to do with entrance and exit effects, where the flow is not viscometric. Very long channels can be used to minimize end effects, but this leads to excessive pressure drops and possible pressure dependence of the viscosity; the use of short capillaries or slits, which is the usual practice, requires a correction for the end effects. Temperature control is a serious problem in all rheological measurements for polymer melts.

The measured viscosities and first normal stress differences for Tsang and Dealy's two polyethylenes are shown in Figure 9.7. The data at shear rates up to $1 \, \mathrm{s}^{-1}$ were obtained in a torsional flow between a cone and plate, while the data at higher shear rates were obtained from the pressure drop in a capillary. The high- and low-shear rate data appear to be consistent, but it is difficult to obtain overlap with a commercial piston-driven capillary viscometer because of the importance of frictional losses at very low rates. These data are typical of many commercial polymers.

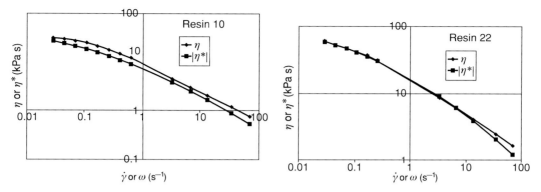

Figure 9.8. Test of the Cox-Merz rule for two polyethylenes: (a) low-density Resin 10 and (b) high-density Resin 22. The curves for $|\eta^*|$ were computed from the spectra in Table 9.2 by P. D. Rai.

Note that the normal stresses can become quite large relative to the shear stresses. (The simplest comparison is at $1\ \mathrm{s}^{-1}$, where the viscosity is numerically equal to the shear stress in consistent SI units.) The lines in the figure refer to a constitutive equation that is discussed subsequently.

9.3.3 Cox-Merz Rule

Linear viscoelastic measurements can provide surprising insight into the steady shear behavior of many polymer melts through an empiricism known as the *Cox-Merz rule*. According to this relation, a plot of $|\eta^*| = [(\eta')^2 + (\eta'')^2]^{1/2}$ versus frequency has the same functionality as a plot of shear viscosity η versus shear rate. The Cox-Merz rule is illustrated for the two polyethylene melts in Figure 9.8. The spectra in Table 9.2 were used to calculate $|\eta^*|$ as a function of frequency, so the comparison is made at the precise rates where the steady shear data exist. The agreement is good for the high-density Resin 22, and not bad for the low-density Resin 10. Good agreement is expected at very low shear rates and frequencies, since $|\eta^*|$ is dominated at low frequencies by η', which must become equal to the zero-shear viscosity η_o in the limit. What is completely unexpected, however, is that the curves are close at *high* frequencies and shear rates; the linear viscoelastic measurement probes the small-amplitude response of an equilibrium structure that is at rest, while the shear viscosity determines the response at very large deformations. The practical significance of this empiricism is profound. Oscillatory shear data can be obtained at much higher frequencies than the shear rates accessible in a torsional geometry in steady flow, especially with the use of time–temperature superposition, and they require much smaller samples. Hence, the Cox-Merz rule is used routinely in many laboratories for the measurement of shear viscosities at rates up to $100\ \mathrm{s}^{-1}$. The theoretical foundations of the Cox-Merz rule are not well established, but recent developments in molecular modeling have begun to unravel the reason for this curious and important phenomenon.

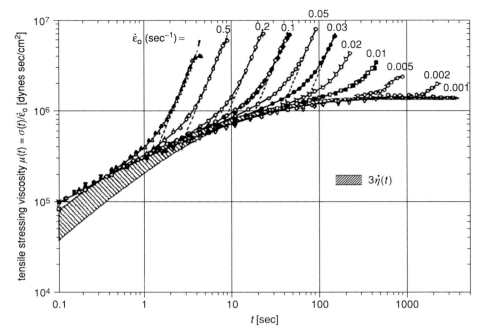

Figure 9.9. Tensile stress divided by stretch rate as a function of time for a low-density polyethylene (Melt I, T = 150 °C). Reprinted with permission from Meissner, *Trans. Soc. Rheol.*, **16**, 405 (1972).

9.4 Extensional Rheology

The description of the kinematics and tensile stress in uniform uniaxial flow was developed in Section 7.2. The experimental design of a system to carry out this flow is very difficult. A number of designs exist, and commercial instruments have been offered, but the measurement is not a routine one and reliable data are limited to a small number of polymer melts. The stretch rates that can be achieved are typically 1 s^{-1} and less.

Figure 9.9 shows a classical data set by Meissner on the low-density polyethylene whose transient shear stress was shown in Figure 2.6. The tensile stress divided by the stretch rate is plotted versus time, together with three times the transient development of the zero-shear viscosity. The data deviate from a single curve at values of the strain (stretch rate multiplied by time) of about 2. There is a plateau at low stretch rates at a Trouton ratio of 3, but the plateau is followed by a sharp increase, and in general the tensile stress greatly exceeds three times the shear viscosity. (The shear viscosity for this polymer decreases with shear rate, so the deviation from three times the viscosity is greater than it would appear when the comparison is based only on the zero-shear viscosity.) The fact that the data lie above the band of three times the shear values at short times is probably an experimental artifact. A steady-state stress is not reached in these experiments, except perhaps at the highest and lowest stretch rates. An apparent steady state has been reported in other measurements.

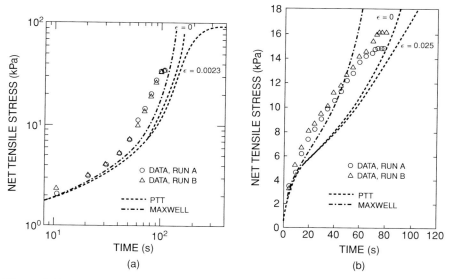

Figure 9.10. Tensile stress as a function of time for two polyethylenes at 170 °C: (a) low-density Resin 10; (b) high-density Resin 22. The lines are the fit to the Phan-Thien/Tanner model, discussed below. Reprinted with permission from Tsang and Dealy, *J. Non-Newtonian Fluid Mech.*, **9**, 203 (1981). Copyright Elsevier.

Tensile stress measurements as a function of time are shown in Figure 9.10 for Tsang and Dealy's two polyethylenes at stretch rates of 0.03 s^{-1}. Results from two runs for each polymer give some idea of the reproducibility. The lines again correspond to constitutive equations to be discussed below. Note that a linear time scale is used for the low-density Resin 10, while the time scale is logarithmic for the high-density Resin 22. There appears to be an approach to a steady-state stress for both polymers. It is interesting to note that the branched resin exhibits a higher tensile stress, whereas it has a lower viscosity than the linear resin.

9.5 Constitutive Equations

9.5.1 Introduction

Equations 9.8 and 9.10b are written for rectilinear flows and infinitesimal deformations. We need equations that apply to finite, three-dimensional deformations. Intuitively, one might expect simply to replace the strain rate $d\gamma/dt$ by the components of the symmetric deformation rate tensor ($\partial v_x/\partial y + \partial v_y/\partial x$, etc.) to obtain a three-dimensional formulation, as in Section 2.2.3, and $d\tau/dt$ by the substantial derivative D/Dt of the appropriate stress components. The first substitution is correct, but intuition would lead us badly astray regarding the second. Constitutive equations must be properly invariant to changes in the frame of reference (they must satisfy the *principle of material frame indifference*), and the substantial derivative of a stress or deformation-rate tensor is not properly invariant. The properly invariant generalization of the derivative of a tensor is known, and it involves nonlinear terms in the components of the stress and the components of the deformation rate. There is, in fact, a three-parameter infinity of such invariant generalizations, but common usage, which we follow here, employs a one-parameter family.

The equivalent problem also arises in the integral equation formulation, Equation 9.13, where it is somewhat easier to appreciate. Here the issue is formulated as the proper definition of strain as measured by an observer in the laboratory frame of reference, and ambiguity can be observed even in one dimension. Consider a material element of length L_o that is stretched an infinitesimal distance dL. The infinitesimal strain is clearly dL/L_o. Now suppose the element is stretched a finite distance, to a length L_1. Selection of the appropriate reference length to use in calculating the strain is no longer obvious. The traditional *engineering strain* uses the initial length as the reference, so the total strain is $(L_1 - L_o)/L_o$. It is reasonable to argue, however, that the strain should be based on the instantaneous length at every step in the deformation; in that case the strain would be the integral of dL/L, or $\ln(L_1/L_o)$. This logarithmic measure of strain is known as *Hencky strain*. Both strain measures are in use, as are others. The issue in three-dimensional constitutive equations is deeper, but not fundamentally different. We will limit our consideration here to the differential equation forms of constitutive equations, so we will not pursue the issue of strain measures further.

9.5.2 Maxwell Equation

The simplest generalization of the linear Equation 9.10 that satisfies the principle of material frame indifference is as follows:

$$\lambda \left[\frac{D\boldsymbol{\tau}}{Dt} - \nabla \mathbf{v}^T \cdot \boldsymbol{\tau} - \nabla \mathbf{v} \cdot \boldsymbol{\tau} \right] + \boldsymbol{\tau} = \lambda G \mathbf{D}. \tag{9.16}$$

We have dropped the subscript n for convenience, but it is implied throughout. $\mathbf{D}$ denotes the terms multiplying the viscosity on the right-hand sides of the entries in Table 2.3. The substantial derivative D/DT equals $\partial/\partial t + v_x \partial/\partial x + v_y \partial/\partial y + v_z \partial/\partial z$ in rectangular Cartesian coordinates; the form in other coordinate systems can be deduced from Table 2.2. The component equations are given in Table 9.3 in rectangular Cartesian coordinates for two-dimensional flow in the xy plane and in cylindrical coordinates for axisymmetric flow with no θ variation or θ component of velocity. (Note that the *hoop* stress $\sigma_{\theta\theta}$ need not vanish in the latter case.) These reduced forms suffice for our purposes. The full three-dimensional equations can be found in texts on rheology. λG(i.e., η) is temperature dependent for nonisothermal flows; λ will usually have the same temperature dependence, since G is insensitive to temperature. Time–temperature superposition also requires a term $\lambda \tau D \ln T/Dt$ on the left side of Equation 9.16, but this term does not generally seem to be important and is usually ignored.

Table 9.3. Components of the Maxwell fluid, Equation 9.16, for two-dimensional flows

Two-dimensional flow in the xy plane

$$xx: \tau_{xx} + \lambda \left[\frac{D\tau_{xx}}{Dt} - 2\frac{\partial v_x}{\partial x}\tau_{xx} - 2\frac{\partial v_x}{\partial y}\tau_{yx} \right] = 2\lambda G \frac{\partial v_x}{\partial x}$$

$$yy: \tau_{yy} + \lambda \left[\frac{D\tau_{yy}}{Dt} - 2\frac{\partial v_y}{\partial x}\tau_{xy} - 2\frac{\partial v_y}{\partial y}\tau_{yy} \right] = 2\lambda G \frac{\partial v_y}{\partial y}$$

$$xy = yx: \tau_{xy} + \lambda \left[\frac{D\tau_{xy}}{Dt} - \frac{\partial v_x}{\partial y}\tau_{yy} - \frac{\partial v_y}{\partial x}\tau_{xx} \right] = \lambda G \left(\frac{\partial v_x}{\partial y} + \frac{\partial v_y}{\partial x} \right)$$

Axisymmetric flow with no θ variation and $v_\theta = 0$

$$rr: \tau_{rr} + \lambda \left[\frac{D\tau_{rr}}{Dt} - 2\frac{\partial v_r}{\partial r}\tau_{rr} - 2\frac{\partial v_r}{\partial z}\tau_{zr} \right] = 2\lambda G \frac{\partial v_r}{\partial r}$$

$$\theta\theta: \tau_{\theta\theta} + \lambda \left[\frac{D\tau_{\theta\theta}}{Dt} - 2\frac{v_r}{r}\tau_{\theta\theta} \right] = 2\lambda G \frac{v_r}{r}$$

$$zz: \tau_{zz} + \lambda \left[\frac{D\tau_{zz}}{Dt} - 2\frac{\partial v_z}{\partial r}\tau_{rz} - 2\frac{\partial v_z}{\partial z}\tau_{zz} \right] = 2\lambda G \frac{\partial v_z}{\partial z}$$

$$rz = zr: \tau_{rz} + \lambda \left[\frac{D\tau_{rz}}{Dt} - \frac{\partial v_r}{\partial z}\tau_{zz} - \frac{\partial v_z}{\partial r}\tau_{rr} \right] = \lambda G \left(\frac{\partial v_r}{\partial z} + \frac{\partial v_z}{\partial r} \right)$$

We first consider the case of a rectilinear flow, in which $v_x = v_x(y)$ and $v_y = v_z = 0$. With $\dot\gamma \equiv dv_x/dy$ the components of Equation 9.16 then simplify to

$$\tau_{xx} - 2\lambda \dot\gamma \, \tau_{yx} = 0, \tag{9.17a}$$

$$\tau_{yy} = 0, \tag{9.17b}$$

$$\tau_{xy} - \lambda \dot\gamma \, \tau_{yy} = \lambda G \dot\gamma . \tag{9.17c}$$

Combination of Equations 9.17b and c then gives, for each value of n,

$$\tau_{n,xy} = \lambda_n G_n \dot\gamma, \tag{9.18a}$$

whereas the nonzero normal stress is then given by

$$\tau_{n,xx} = 2\lambda_n^2 G_n \dot\gamma^2 . \tag{9.18b}$$

The shear rate terms factor out in the summation over all modes, so we obtain, finally,

$$\tau_{xy} = \left(\sum_n \lambda_n G_n \right) \dot\gamma, \tag{9.19a}$$

$$N_1 = \sum_n (\tau_{n,xx} - \tau_{n,yy}) = 2 \left(\sum_n \lambda_n^2 G_n \right) \dot\gamma^2 . \tag{9.19b}$$

The Maxwell equation with one or multiple modes therefore predicts a shear stress that varies linearly with shear rate and a finite first normal stress difference that varies quadratically with shear rate. The second normal stress difference is predicted to be zero. The linear/quadratic behavior is consistent with steady shear behavior at low shear rates, but it does not correctly reflect the shear thinning at high rates. The shear viscosity η equals $\sum \lambda_n G_n$, which is the limiting value of η' at low frequency, as required (cf. Equation 9.11a). The coefficient of the quadratic term in the normal stress is $2\sum \lambda_n^2 G_n$, which equals twice the coefficient of the quadratic term in G' at low frequencies (Equation 9.11b). A mean relaxation time is often defined as

$$\lambda = \frac{\sum \lambda_n^2 G_n}{\sum \lambda_n G_n} = \frac{N_1}{2\tau_{xy}\,\dot{\gamma}}. \tag{9.20}$$

We now turn to uniform uniaxial extensional flow, with (cf. Equation 7.4a–b) the velocity field $v_z = \dot{\gamma}_E\, z$, $v_r = -\frac{1}{2}\dot{\gamma}_E\, r$. The velocity gradients are all constant in space, so it follows that the stresses are independent of spatial position. It is easily shown with this velocity field that $\tau_{rz} = 0$ and $\tau_{\theta\theta} = \tau_{rr}$. The total axial stress σ_{zz} is then equal to the stress difference $\tau_{zz} - \tau_{rr}$ (cf. Equation 7.9). The component equations for each dynamical mode are

$$\tau_{rr} + \lambda \left[\frac{\partial \tau_{rr}}{\partial t} + \dot{\gamma}_E\, \tau_{rr} \right] = -\lambda G\, \dot{\gamma}_E, \tag{9.21a}$$

$$\tau_{zz} + \lambda \left[\frac{\partial \tau_{zz}}{\partial t} - 2\dot{\gamma}_E\, \tau_{zz} \right] = 2\lambda G\, \dot{\gamma}_E. \tag{9.21b}$$

We assume a constant stretch rate, in which case these are simply uncoupled first-order linear ordinary differential equations, which are easily integrated. When the modes are summed to give the total stress, we then obtain the equation

$$\sigma_{zz} = 3\left(\sum_n \frac{\lambda_n G_n}{\left(1 - 2\lambda_n \dot{\gamma}_E\right)\left(1 + \lambda_n \dot{\gamma}_E\right)} \right)\dot{\gamma}_E - \sum_n \frac{2\lambda_n G_n \dot{\gamma}_E}{1 - 2\lambda_n \dot{\gamma}_E} e^{-(1 - 2\lambda_n \dot{\gamma}_E)t/\lambda_n}$$

$$- \sum_n \frac{\lambda_n G_n \dot{\gamma}_E}{1 + \lambda_n \dot{\gamma}_E} e^{(1 + \lambda_n \dot{\gamma}_E)t/\lambda_n}. \tag{9.22}$$

The dynamical behavior is striking. A steady-state stress is reached as long as $2\lambda_n \dot{\gamma}_E < 1$ for all modes; the Trouton ratio of 3 is attained at steady state when $2\lambda_n \dot{\gamma}_E \ll 1$ for all modes, while for all finite values of this product the ultimate steady-state Trouton ratio is greater than 3. If $2\lambda_n \dot{\gamma}_E > 1$ for any single mode – clearly, the longest relaxation time is the one that matters here – the stress will ultimately grow without bound. The response for any single mode is shown in Figure 9.11. The response for two modes, using mean parameters characteristic of a polyethylene, is shown in Figure 9.12. The mean relaxation time is defined as in Equation 9.20. Even two dynamical modes are sufficient to permit a plateau (seen here as the inflection) followed by a second region of increasing stress, but an ultimate steady state will not be achieved with this simple model.

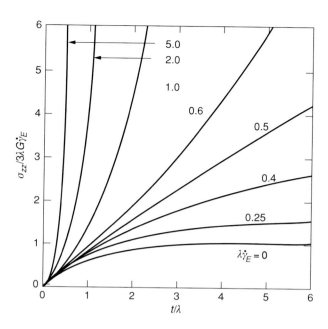

Figure 9.11. Tensile stress divided by $3\lambda G\,\dot{\gamma}_E$ as a function of reduced time for a single mode of a Maxwell fluid. Reprinted with permission from Denn and Marrucci, *AIChE J.*, **17**, 101 (1971). Copyright American Institute of Chemical Engineers.

9.5.3 Phan-Thien/Tanner Model

The *Phan-Thien/Tanner* (PTT) model is one of many generalizations that have been introduced to deal with the deficiencies of the basic Maxwell model: constant viscosity, quadratic first normal stress difference, zero second normal stress difference, and infinite tensile stress at a finite extension rate. Many of these models, including the PTT, are derived from microstructural models that attempt to account for aspects of chain morphology and interactions. PTT is a *network* model, in which the chains are assumed to interact at entanglement points. There are kinetic expressions

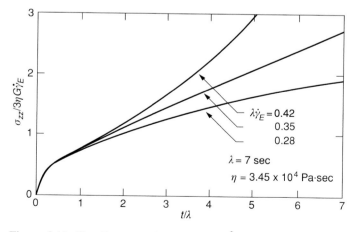

Figure 9.12. Tensile stress divided by $3\eta\,\dot{\gamma}_E$ as a function of reduced time for a two-mode Maxwell fluid. η is the viscosity and λ is the mean relaxation time. Reprinted from Denn, in R. S. Rivlin, ed., *The Mechanics of Viscoelastic Fluids*, AMD Vol. 22, ASME, New York, 1977, p. 101.

Table 9.4. Components of the Phan-Thien/Tanner Fluid, Equation 9.23, for two-dimensional flows

Two-dimensional flow in the xy plane

$$xx:\ Y\tau_{xx} + \lambda\left[\frac{D\tau_{xx}}{Dt} - 2\frac{\partial v_x}{\partial x}\tau_{xx} - 2\frac{\partial v_x}{\partial y}\tau_{yx} + 2\xi\frac{\partial v_x}{\partial x}\tau_{xx} + \xi\left(\frac{\partial v_x}{\partial y} + \frac{\partial v_y}{\partial x}\right)\tau_{xy}\right] = 2\lambda G\frac{\partial v_x}{\partial x}$$

$$yy:\ Y\tau_{yy} + \lambda\left[\frac{D\tau_{yy}}{Dt} - 2\frac{\partial v_y}{\partial x}\tau_{xy} - 2\frac{\partial v_y}{\partial y}\tau_{yy} + 2\xi\frac{\partial v_y}{\partial y}\tau_{yy} + \xi\left(\frac{\partial v_x}{\partial y} + \frac{\partial v_y}{\partial x}\right)\tau_{xy}\right] = 2\lambda G\frac{\partial v_y}{\partial y}$$

$$xy = yx:\ Y\tau_{xy} + \lambda\left[\frac{D\tau_{xy}}{Dt} - \frac{\partial v_x}{\partial y}\tau_{yy} - \frac{\partial v_y}{\partial x}\tau_{xx} + \frac{1}{2}\xi\left(\frac{\partial v_x}{\partial y} + \frac{\partial v_y}{\partial x}\right)(\tau_{xx} + \tau_{yy})\right] = \lambda G\left(\frac{\partial v_x}{\partial y} + \frac{\partial v_y}{\partial x}\right)$$

$$Y = \exp\left[\varepsilon(\tau_{xx} + \tau_{yy} + \tau_{zz})/G\right]$$

Axisymmetric flow with no θ variation and $v_\theta = 0$

$$rr:\ Y\tau_{rr} + \lambda\left[\frac{D\tau_{rr}}{Dt} - 2\frac{\partial v_r}{\partial r}\tau_{rr} - 2\frac{\partial v_r}{\partial z}\tau_{zr} + 2\xi\frac{\partial v_r}{\partial r}\tau_{rr} + \xi\left(\frac{\partial v_r}{\partial z} + \frac{\partial v_z}{\partial r}\right)\tau_{rz}\right] = 2\lambda G\frac{\partial v_r}{\partial r}$$

$$\theta\theta:\ Y\tau_{\theta\theta} + \lambda\left[\frac{D\tau_{\theta\theta}}{Dt} - 2\frac{v_r}{r}\tau_{\theta\theta} + 2\xi\frac{v_r}{r}\tau_{\theta\theta}\right] = 2\lambda G\frac{v_r}{r}$$

$$zz:\ Y\tau_{zz} + \lambda\left[\frac{D\tau_{zz}}{Dt} - 2\frac{\partial v_z}{\partial r}\tau_{rz} - 2\frac{\partial v_z}{\partial z}\tau_{zz} + 2\xi\frac{\partial v_z}{\partial z}\tau_{zz} + \xi\left(\frac{\partial v_r}{\partial z} + \frac{\partial v_z}{\partial r}\right)\tau_{rz}\right] = 2\lambda G\frac{\partial v_z}{\partial z}$$

$$rz = zr:\ Y\tau_{rz} + \lambda\left[\frac{D\tau_{rz}}{Dt} - \frac{\partial v_r}{\partial z}\tau_{zz} - \frac{\partial v_z}{\partial r}\tau_{rr} - \frac{v_r}{r}\tau_{rz} + \xi\left(\frac{\partial v_r}{\partial z} + \frac{\partial v_z}{\partial r}\right)(\tau_{rr} + \tau_{zz}) - 2\xi\frac{v_r}{r}\tau_{rz}\right] = \lambda G\left(\frac{\partial v_z}{\partial r} + \frac{\partial v_r}{\partial z}\right)$$

$$Y = \exp\left[\varepsilon(\tau_{rr} + \tau_{\theta\theta} + \tau_{zz})/G\right]$$

describing the rate of formation and disappearance of entanglements; the former is driven by the distance from equilibrium, while the latter depends on the stress. (The Maxwell equation is obtained rigorously from such a kinetic model when both rates are independent of the deformation.) There are, of course, approximations and empiricisms introduced in passing from the basic network formulation to the final constitutive equation. The model, which contains only two additional parameters beyond those measured in a linear viscoelastic experiment, has been relatively successful in describing melt rheology and it has been used extensively in simulations of polymer processing flows. Each term τ_n in the PTT equation satisfies the following Maxwell-like equation:

$$\lambda\left[\frac{D\tau}{Dt} - \nabla\mathbf{v}^T\cdot\tau - \tau\cdot\nabla\mathbf{v} + \xi(\tau\cdot\mathbf{D} + \mathbf{D}\cdot\tau)\right] + Y(tr\tau/G)\,\tau = \lambda G\mathbf{D}. \qquad (9.23)$$

The components for two-dimensional flows are shown in Table 9.4. $tr\tau$ is the *trace* of the stress; the trace is the sum of the diagonal components and is a scalar invariant (the *first invariant*) that is the same in all coordinate systems. Two forms have commonly been used for the function Y:

$$Y(tr\tau/G) = \exp(\varepsilon tr\tau/G), \qquad (9.24a)$$

$$Y(tr\tau/G) = 1 + \varepsilon tr\tau/G. \qquad (9.24b)$$

The exponential form is preferred for polymer melts and has been used in all the examples cited here. The function Y plays an important role in extensional flows, where it prevents the stress from growing without bound for finite stretch rates.

(The structure is essentially that of a Maxwell fluid if Equation 9.23 is multiplied by Y^{-1}, but with a relaxation time equal to λY^{-1} that decreases with increasing stress.) Values of ε are typically of order 0.02 or less, and at this level the shear viscosity and normal stress differences are insensitive to ε. The parameter ξ arises in the network model as a *slip coefficient* that reflects the motion of the network relative to the continuum.

The equations for the PTT fluid do not have a closed-form analytical solution for a rectilinear flow $v_x = v_x(y)$, $v_y = v_z = 0$. To a good approximation, however, we can take Y equal to unity in a shear flow, in which case the viscometric functions are obtained approximately as follows:

$$\tau_{xy} = \sum_n \frac{\lambda_n G_n \dot{\gamma}}{1 + \xi (2 - \xi) \left(\lambda_n \dot{\gamma}\right)^2}, \tag{9.25a}$$

$$N_1 = \tau_{xx} - \tau_{yy} = \sum_n \frac{2\lambda_n^2 G_n \dot{\gamma}^2}{1 + \xi (2 - \xi) \left(\lambda_n \dot{\gamma}\right)^2}, \tag{9.25b}$$

$$N_2 = \tau_{yy} - \tau_{zz} = -\tfrac{1}{2}\xi N_1. \tag{9.25c}$$

Thus, the fluid is shear thinning for $0 < \xi < 2$, with a finite second normal stress difference. $-N_2/N_1$ is typically of order 0.1, so ξ will typically be of order 0.2. There is no simple analytical solution to the equations for transient stress growth in uniform uniaxial extension, and the coupled ordinary differential equations must be solved numerically.

Phan-Thien and Tanner used data of Meissner for the low-density polyethylene in Figure 9.9 to test the model, with $\xi = 0.2$ and $\varepsilon = 0.01$. The fit to steady shear viscosity and normal stresses is good, and the transients are fit reasonably well; the shear data are insensitive to the value of ε as expected. The fit to the extensional data at three stretch rates is shown in Figure 9.13. The agreement is quite good, especially when it is recalled that there is probably an experimental artifact at short times that causes the data to be high. The model predicts an approach to a steady extensional viscosity that scales as $1/\varepsilon$, but the data do not extend sufficiently far to test the prediction.

The fit of the PTT equation to the steady shear data of Tsang and Dealy is shown as the lines in Figures 9.7, with $\xi = 0.1$ for LDPE Resin 10 and $\xi = 0.3$ for HDPE Resin 22. The fits are reasonable, given that the nonlinearity in the shear data is accounted for by only a single parameter, but the model predicts shear thinning that is more rapid than found in the data, especially for the HDPE. The fit to the extensional data at a stretch rate of 0.03 s^{-1} is shown in Figure 9.10. The fit is much poorer than that found by Phan-Thien and Tanner for the Meissner data, and the apparent steady state is attained earlier and at a lower stress than predicted by the equation. The Maxwell model, which corresponds to $\xi = \varepsilon = 0$, is shown for comparison. The Maxwell model does a better job of fitting these extensional data, but of course the

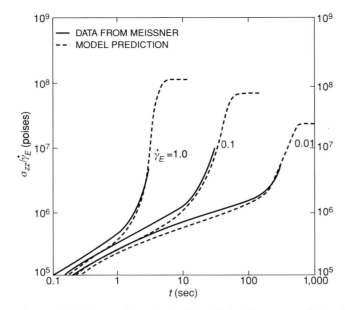

Figure 9.13. Comparison of the Phan-Thien/Tanner model to the extensional data of Meissner. Reprinted with permission from Phan-Thien and Tanner, *J. Non-Newtonian Fluid Mech.*, **2**, 353 (1977). Copyright Elsevier.

Maxwell model cannot account for the variable viscosity and nonquadratic normal stress.

9.5.4 Other Constitutive Equations

There are numerous other constitutive equations of both differential and integral type for polymer melts, and some do a better job of matching data from a variety of experiments than does the PTT equation. The overall structure of the differential equations is usually of the form employed here: The total stress is a sum of individual stress modes, each associated with one term in the linear viscoelastic spectrum, and there is an invariant derivative similar in structure to the one in the PTT equation, but with different quadratic nonlinearities in τ and $\nabla\mathbf{v}$. The Giesekus model, for example, which is also widely used, has the following form:

$$\lambda\left[\frac{D\tau}{Dt} - \nabla\mathbf{v}^T\cdot\tau - \tau\cdot\nabla\mathbf{v} + \frac{\alpha}{\lambda G}(\tau\cdot\tau)\right] + \tau = \lambda G\mathbf{D}, \quad 0 \le \alpha \le 1. \tag{9.26}$$

This model also has a shear-thinning viscosity, a finite second normal stress difference, and bounded extensional stresses.

The modern approach to constitutive equations for melts and concentrated solutions has evolved from the *reptation* idea popularized by de Gennes and Doi and Edwards. In this picture, the mobility of a polymer chain in any direction except along the backbone is visualized as constrained to an imaginary tubelike region formed by the entanglements with the neighboring chains, which serve to restrict any transverse motion. This concept leads to an expression for the linear viscoelastic spectrum of linear chains, a particular strain measure, and, with approximations, an explicit constitutive equation that can be closely approximated by a differential equation of the generalized Maxwell type. The advantage of this formulation is that molecular weight scaling is incorporated directly into the coefficients. The simple reptation idea is powerful, but the equation that results is too simple and sometimes aphysical, so considerable effort has been devoted to improved formulations of tube models. The tube approach has enabled the development of models that account for molecular weight distributions, and the concepts can be extended to include branching. The *pom-pom model*, a tube model that was developed for branched polymers but also works well for linear polymers, can be expressed in differential equation form and seems to be quite versatile. It has been used for some processing simulations and has become a component of some commercial codes. Passing from the PTT model to one of these improved differential constitutive equations is a matter of programming detail rather than new basic concepts, and we will not pursue them further.

Invariant integral generalizations of Equation 9.14 lead to a class of constitutive equations known as BKZ or K-BKZ (for Kaye, Bernstein, Kearsley, and Zapas) models. This class includes most of the equations derived from the tube concept. A rigorous one-to-one correspondence between integral and differential constitutive equations exists only for the basic Maxwell fluid, but good differential equation approximations exist for many of the tube models. Implementation of an integral constitutive equation requires exact tracking of the strain for each fluid element, which involves very different programming considerations from simultaneous solution of the differential equation models with the momentum and energy equations. The advantage of the integral formulation is that the computational intensity is independent of the number of modes in the memory function, but this advantage is balanced by the more difficult implementation.

9.6 Entry and Exit Losses

Viscosity determination using pressure-driven flow in a capillary or slit is confounded by losses in the entry and exit regions, where the flow is readjusting and the velocity profile is not fully developed. It is possible to use very long capillaries or slits in order to minimize this effect, but the resulting large pressure drops can then introduce problems associated with the pressure dependence of the viscosity, as well as viscous dissipation.

The excess pressure drop associated with the entrance and exit is shear-rate dependent for polymer melts. If the excess pressure drop can be measured, then it can be subtracted from the total pressure drop to provide the value for fully developed flow, and the latter can be used to compute the shear stress at a given shear rate. The method commonly used is known as a *Bagley plot*. Here, the pressure drop as a function of shear rate is measured in capillaries of varying length-to-diameter (L/D) ratios. The data are then plotted as pressure drop versus L/D at fixed shear

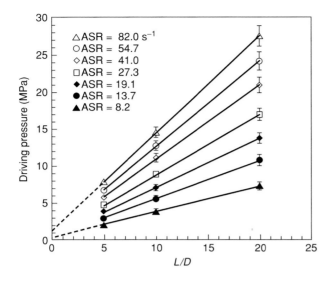

Figure 9.14. Bagley plot for a linear metallocene polyethylene at 150 °C. "ASR" denotes "apparent shear rate." Reprinted with permission from Kim and Dealy, *J. Rheol.*, **45**, 1413 (2001).

rate and extrapolated to $L/D = 0$. An example by Kim and Dealy is shown in Figure 9.14 for a linear metallocene polyethylene with $M_w = 109{,}000$ and $M_w/M_n = 2.3$. Only the highest and lowest shear rates are extrapolated here for clarity. Data for $L/D = 40$, which results in a nonlinear curve at each shear rate because of pressure effects, are not included. Kim and Dealy showed that the same result could be obtained using a properly designed capillary with $L/D < 0.5$ (an *orifice die*). Standard practice in many laboratories is to use a capillary with L/D of 10 or 15 and an orifice die, and then to subtract the pressure drop through the orifice from the total pressure drop in the longer die to obtain the fully developed value.

The entry and exit losses for a Newtonian fluid at very low Reynolds number are equivalent to fully developed flow in an extra capillary length of about one diameter. The extra losses can be significantly larger for polymer melts, however. Many authors believe that the large entry and exit losses for polymer melts are a consequence of the stretching flow along the centerline of the entry region, and approximate analyses by Cogswell and Binding are often used to infer extensional viscosities from the entry losses. These analyses involve some rather severe approximations and assumptions, and direct comparisons with uniform uniaxial extensional measurements in the same range of deformation rates are rare. It can be shown rigorously for slits that the first order correction to the losses through a contraction resulting from fluid elasticity is *negative* (although this first-order regime is unattainable in practice and negative values are never observed), in which case the Cogswell and Binding analyses would give a thermodynamically impossible negative extensional viscosity.

Extrapolation of the pressure profile in a slit right up to the exit sometimes results in a positive nonzero value. If fully developed flow were to exist right up to the exit, it could then be shown that the first normal stress difference could be extracted from this exit pressure, and the method has been used. It is easy to show, however, that the error resulting from flow rearrangement near the exit is larger than the normal stress, and the uncertainty in the extrapolation of the pressure also introduces an error that is larger than the normal stress. Hence, data obtained using this method should not be used.

9.7 Concluding Remarks

The Phan-Thien/Tanner constitutive equation does not represent the state of the art in modeling melt flow at the time of this writing, but it is adequate to illustrate the response of melts of flexible polymers in complex flows and it has a mathematical structure that does not differ substantively from other equations with a firmer basis in molecular theory. Furthermore, it has been widely used in simulation studies to date. Hence, we will use it for illustrative purposes in this text, recognizing that it is likely to be replaced as the preferred constitutive equation for applications. The minimum rheological information required for simulations is thus the temperature-dependent linear viscoelastic spectrum and the temperature-dependent viscosity as a function of shear rate. Extensional data should be used, but they are often unavailable; when the PTT equation is employed it is therefore common to select a "reasonable" value of ε to describe the extensional response.

Finally, it is important to note that our focus in this chapter and throughout the text is on single-phase polymers with flexible backbones. The rheology of polymers with rigid elements in the backbone, which are often liquid crystalline in the melt, is quite different from that described here. Immiscible blends and filled polymers may have behavior similar to that of the flexible melts, but there are often important differences. Structured fluids are discussed briefly in Chapter 13.

BIBLIOGRAPHICAL NOTES

Basic textbooks covering polymer melt rheology include

Dealy, J. M., and R. G. Larson, *Structure and Rheology of Molten Polymers*, Hanser, New York, 2006.
Dealy, J. M., and K. F. Wissbrun, *Melt Rheology and Its Role in Plastics Processing: Theory and Applications*, Van Nostrand Reinhold, New York, 1990.
Gupta, R. K., *Polymer and Composite Rheology*, 2nd ed., Marcel Dekker, New York, 2000.
Larson, R. G., *Constitutive Equations for Polymer Melts and Solutions*, Butterworth, Boston, 1988.
Larson, R. G., *The Structure and Rheology of Complex Fluids*, Oxford, New York, 1999.
Macosko, C. W., *Rheology: Principles, Measurements, and Applications*, VCH, New York, 1994.
Morrison, F. A., *Understanding Rheology*, Oxford, New York, 2001.
Tanner, R. I., *Engineering Rheology*, 2nd ed., Oxford, New York, 2000.

More recent approaches to extensional measurements on melts are in

Bach, A., H. K. Rasmussen, and O. Hasager, *J. Rheol.*, **47**, 429 (2003).
Sentmanat, M., B. N. Wang, and G. H. McKinley, *J. Rheol.*, **49**, 585 (2005).

All of the books listed above are adequate for the material covered in this chapter and needed for the remainder of this text. The most recent books are the most up to date on the molecular-based theories.

Tube models are reviewed in

McLeish, T. C. B., *Adv. Phys.*, **51**, 1379 (2002).

The pom-pom model was introduced by McLeish and Larson in 1998, but a slight variant known as the *extended pom-pom model* (XPP) is more commonly used in simulations; see

Verbeeten, W. M. H., Peters, G. W. M., and Baaijens, F. P. T., *J. Rheol.*, **45**, 823 (2001) [Erratum, **45**, 1489 (2001)].

The extended pom-pom model sometimes predicts unphysical behavior, as discussed in

Inkson, N. J., and T. N. Philllips, *J. Non-Newtonian Fluid Mech.*, **145**, 92 (2007).

The literature on this subject is growing, and the best sources are articles in the *Journal of Rheology*, *Rheologica Acta*, the *Journal of Non-Newtonian Fluid Mechanics*, and *Macromolecules*.

10 Viscoelasticity in Processing Flows

10.1 Introduction

Viscoelasticity will clearly have a large effect in some processing operations and little or none in others, and we require a way to discriminate between these cases. One clue follows from the linear viscoelastic experiments shown in Figures 9.2 and 9.3 and the accompanying spectral description in Equations 9.11a–b. The entangled network is able to relax at low frequencies, so the elastic contribution to the stress is negligible and the deformation is mostly dissipative ($G' \to 0$). The stress at high frequencies cannot relax, so dissipation is negligible and the deformation is recoverable ($\eta' \to 0$). The transition between these two extremes is sharp for a liquid with a single Maxwell mode and occurs in the neighborhood of $\lambda\omega \sim 1$. ω^{-1} is the characteristic time for the oscillatory deformation, so we may think of the two limiting cases as representing processes that are slow and fast, respectively, relative to the characteristic time of the fluid. The transition is murkier for most polymer melts, where there are many dynamical modes, but there will be some relaxation time – a mean value like that given by Equation 9.20 or the longest relaxation time in the spectrum – such that the same criterion can be usefully applied. The ratio of the characteristic time of the fluid to the characteristic time of the process is known as the *Deborah number*[*] and is usually denoted *De*. The time scale for the process is usually the residence time. Thus, extrusion, which has a very long residence time relative to the relaxation time of the melt, is a low Deborah number process and should be dominated by dissipative effects. Melt spinning, in which the residence time in the melt zone is very short, may be a high Deborah number process in which elasticity will dominate.

Another useful dimensionless measure follows from a traditional dimensional analysis approach. Consider the Maxwell equation,

$$\lambda \left[\frac{D\boldsymbol{\tau}}{Dt} - \nabla\mathbf{v}^T \cdot \boldsymbol{\tau} - \boldsymbol{\tau} \cdot \nabla\mathbf{v} \right] + \boldsymbol{\tau} = \lambda G\mathbf{D}. \tag{10.1}$$

[*] From the Song of Deborah, Judges 5:5, "The mountains quaked (sometimes translated as "flowed") at the presence of the Lord." The concept of different types of deformation on different time scales and the name of the dimensionless group were introduced by Marcus Reiner in 1964, although Reiner's definition of *De* was the inverse of the definition now in use.

We suppose that there is only one characteristic length in the system, L, and one characteristic velocity, V. We further assume that the only characteristic time is L/V. We then define dimensionless quantities, denoted by an asterisk (*), in the usual way:

$$t = t^*L/V, \quad v = v^*V, \quad \nabla = \nabla^*/L, \quad \mathbf{D} = \mathbf{D}^*V/L.$$

There are two possible choices of the characteristic stress, G and $\lambda GV/L$. The former is the modulus, while the latter is proportional to the shear stress. (Recall that the viscosity $\eta = \lambda G$ for a Maxwell fluid.) We choose to normalize with the shear stress, so

$$\boldsymbol{\tau} = \boldsymbol{\tau}^*\lambda GV/L.$$

Substitution of these new variables into Equation 10.1 then leads immediately to the dimensionless equation

$$\lambda \frac{V}{L}\left[\frac{D\boldsymbol{\tau}^*}{Dt^*} - \nabla^*\mathbf{v}^{*T}\cdot\boldsymbol{\tau}^* - \boldsymbol{\tau}^*\cdot\nabla^*\mathbf{v}^*\right] + \boldsymbol{\tau}^* = \mathbf{D}^*. \tag{10.2}$$

There is a single dimensionless group, $\lambda V/L$, which is known as the *Weissenberg number*, denoted by various authors as *We* or *Wi*. (*We* is more common, but it can lead to confusion with the Weber number, so *Wi* will be used here.) The shear rate in any viscometric flow is equal to a constant multiplied by V/L, so it readily follows that the ratio of the first normal stress difference to the shear stress is equal to twice that constant multiplied by *Wi*. Hence, *Wi* can be interpreted as the relative magnitude of elastic (normal) stresses to shear stresses in a viscometric flow. The ratio of the shear stress to the shear modulus, G, is sometimes known as the *recoverable shear* and is denoted S_R. S_R differs from *Wi* for a Maxwell fluid only by the constant that multiplies V/L to form the shear rate for a given flow. In fact, many authors define *Wi* as the product of the relaxation time and the shear rate, in which case $Wi = S_R$. It is important to keep the various definitions of *Wi* in mind when comparing results from different authors.

The residence time in many flows is of order L/V (flow past a sphere, for example), in which case *De* and *Wi* are the same. This has caused some confusion, and there are papers in the literature where *De* is used in place of *Wi*. Both groups will appear naturally, and are independent, in flows in which there are two characteristic length scales, one defining local shear stresses and the other defining the residence time; consider flow in a channel, for example, where *De* is usually small while *Wi* can be large.

10.2 Extrudate Swell

As a first example of elastic effects, consider a low Reynolds number jet emerging from a cylindrical tube. We assume that the tube is sufficiently long that the flow is fully developed prior to the exit ($De = 0$). It is well known – indeed, it is a standard problem in first courses in fluid mechanics – that the diameter of a jet of a

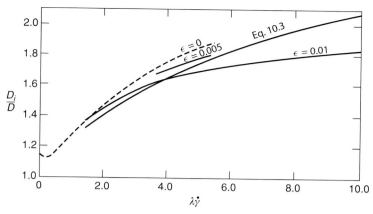

Figure 10.1. Computed extrudate swell of a Phan-Thien/Tanner fluid with $\xi = 0$ for various values of ε. $\varepsilon = 0$ corresponds to a Maxwell fluid. Reprinted by permission of Oxford University Press from Tanner, *Engineering Rheology*, 2nd ed., Oxford, New York, 2000, p. 429.

Newtonian fluid is different from the tube diameter because of the rearrangement from a parabolic laminar velocity profile in the tube to a flat profile in the developed jet. The standard result in fluid mechanics texts is that the jet-to-tube diameter ratio, D_j/D, is 0.82. This is a high Reynolds number result, however. It is less well known that there is a transition in the neighborhood of $Re = 50$ and that the ratio D_j/D for *low Re* is 1.13; that is, the low Reynolds number jet *swells* upon exiting, as shown in Figure 1.10.

We should expect even greater extrudate swell for a viscoelastic liquid since the first normal stress difference "pushes" against the walls of the tube until the exit, after which the resistance vanishes. Tanner has developed an approximate theory for extrudate swell based on the elastic recovery when the tube wall is instantly removed, giving the following result:

$$\frac{D_j}{D} = 0.13 + \left[1 + \frac{1}{8} \left(\frac{N_1}{\tau} \right)^2 \right]^{1/6}. \tag{10.3}$$

τ is the shear stress, and the stresses are evaluated at the shear rate at the wall, which is $8V/D$ for a fluid with a constant viscosity. The additive factor 0.13 is an empirical addition to give the correct result for a Newtonian fluid. Equation 10.3 provides a rough fit to experimental data for polymer melts and concentrated solutions.

Figure 10.1 shows numerical calculations of extrudate swell for a single-mode Phan-Thien/Tanner equation with $\xi = 0$ and various values of ε, together with Equation 10.3. The dashed curve for $\varepsilon = 0$ corresponds to a Maxwell fluid. Note the small, unexpected *decrease* in the extrudate swell for very small values of Wi. Extrudate swell is clearly very important, and swell needs to be taken into account in relating the shape of an extrudate to the die from which it emerges. The problem of designing the shape of a die for an extrudate of complex shape requires a three-dimensional calculation and a constitutive equation that is a very good representation of the polymer properties. Some progress has been made on this problem, but practical calculations would require significant computational resources.

The curves for $\varepsilon = 0$ and 0.005 terminate before the curve for $\varepsilon = 0.01$ because the numerical scheme failed to converge. (These calculations were done using a boundary element method rather than a finite element method, but the results from the latter are essentially the same, as they are for finite difference methods.) The reason for the convergence failure is the stress singularity at the point of departure from the die. The stress in a Newtonian fluid increases roughly like the $-1/2$ power of distance from the singular point, and most numerical schemes are sufficiently robust to deal with this very rapid change over a small distance. The strength of the singularity is not known for viscoelastic fluids, but at least for the Maxwell fluid it appears to be much stronger than for the Newtonian fluid, and all computational schemes ultimately fail at some throughput because of the development of a (numerical) stress boundary layer that propagates through the flow field and contaminates the calculations. The more shear thinning the fluid, the greater the range of flow rates that can be accommodated. This observation can be understood in terms of the PTT equation with $\xi = 0$ by noting that the equation is equivalent to a Maxwell fluid with a relaxation time equal to $\lambda \exp(-\varepsilon tr \tau/G)$; hence, the higher the stress, the smaller the relaxation time and the smaller the effective Weissenberg number. The *high Weissenberg number problem* in viscoelastic computation always arises, and while there has been considerable progress on extending the range of convergence, the fundamental problem is unsolved and convergence will fail at some value of *Wi* for this flow and other flows with singularities.

10.3 Fiber Spinning Revisited

Fiber spinning, discussed in Chapter 7, is a process in which the residence time is short; hence, transient viscoelastic effects are likely to be important. Our starting point for a steady-state thin filament analysis is Equation 7.26:

$$w\frac{dv}{dz} = \frac{d}{dz}\left[\pi R^2 \left(\tau_{zz} - \tau_{rr}\right)\right] - \pi R \rho_a v^2 c_D + \pi \rho R^2 g. \tag{10.4}$$

Here we have removed the overbars from the averaged quantities and made use of the fact that $\tau_{rr} = \tau_{\theta\theta}$. (This equality does not hold for hollow-fiber spinning.) The steady-state equations for each mode of a Phan-Thien/Tanner fluid (Table 9.4) are as follows:

$$Y\tau_{zz} + \lambda\left[v\frac{d\tau_{zz}}{dz} - 2(1-\xi)\frac{dv}{dz}\tau_{zz}\right] = 2\lambda G\frac{dv}{dz}, \tag{10.5a}$$

$$Y\tau_{rr} + \lambda\left[v\frac{d\tau_{rr}}{dz} - (1-\xi)\frac{dv}{dz}\tau_{rr}\right] = -\lambda G\frac{dv}{dz}, \tag{10.5b}$$

$$Y = \exp\left[\varepsilon(\tau_{zz} + 2\tau_{rr})/G\right]. \tag{10.5c}$$

The Maxwell fluid corresponds to $\xi = \varepsilon = 0$.

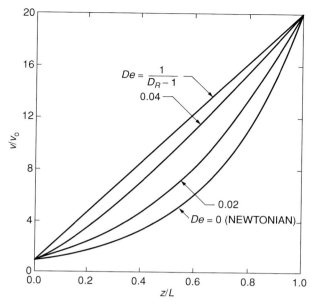

Figure 10.2. Velocity profile in a Maxwell fluid with $\tau_{rr}/\tau_{zz} = 0$ at $z = 0$. Reprinted with permission from Denn et al., *AIChE J.*, **21**, 791 (1975). Copyright American Institute of Chemical Engineers.

An analytical solution to these equations is not possible, but asymptotic solutions in the absence of heat transfer, inertia, air drag, and gravity can be obtained for the single-mode Maxwell fluid in small- and large-force limits. The latter is of the most interest to us here and is written in terms of two dimensionless groups, $De = \lambda v_o/L$ and $B = 4F/3\pi\lambda Gd_o v_o$. v_o and d_o are the initial velocity and diameter, respectively. F is the force, which is a constant along the entire spinline in this limit. The velocity profile is as follows:

$$\frac{v}{v_o} = 1 + \frac{z}{LDe} - \frac{1}{2BDe}\frac{d_o}{L}\ln\left(1 + \frac{z}{LDe}\right)$$

$$- \frac{1}{2}\left[\frac{1}{2BDe}\frac{d_o}{L} + 3\frac{\tau_{rr}(0)}{\tau_{zz}(0)}\right]\left\{1 - \frac{1}{\left(1 + \frac{z}{LDe}\right)^2}\right\} + \cdots . \qquad (10.6)$$

Note that the natural length scale LDe is λv_o. The grouping $BDeL/d_o$ is simply the initial spinline stress divided by the tensile (Young's) modulus, $3G$. The initial ratio of the radial to axial extra stress, $\tau_{rr}(0)/\tau_{zz}(0)$, is required. As $F \to \infty$ ($B \to \infty$) the velocity profile approaches the straight line $1 + z/LDe$, and there is a maximum achievable draw ratio $D_R = v_L/v_o = 1 + De^{-1}$. This result is the analog of the infinite force in a Maxwell fluid at a finite rate of extension.

The solution to the thin filament equations for an isothermal Maxwell fluid in the absence of inertia, air drag, and gravity is shown in Figure 10.2 for $D_R = 20$. The calculation is for the initial value of the ratio τ_{rr}/τ_{zz} set to 0, but the stress ratio approaches 0 on the spinline for all choices in the permissible range from 0

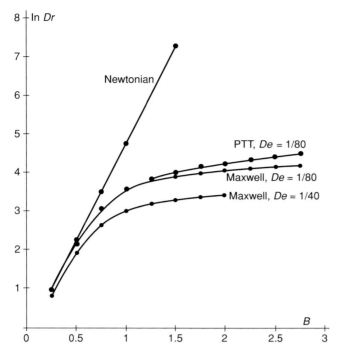

Figure 10.3. Draw ratio at $z = 5d_o$ as a function of dimensionless force for Maxwell and Phan-Thien/Tanner fluids. Reprinted with permission from Keunings et al., *Ind. Eng. Chem. Fundam.*, **22**, 347 (1983). Copyright American Chemical Society.

to $-1/2$ for finite De, and the result is insensitive to the initial value. The approach to a linear profile is clearly seen as De approaches the limiting value defined by $(D_R - 1)^{-1} = 1/19 = 0.0526\ldots$.

Results of a finite element solution without inertia, air drag, or gravity are shown in Figure 10.3 for the isothermal Maxwell and Phan-Thien/Tanner equations, the latter with $\xi = 0$ and $\varepsilon = 0.015$. The computational domain included the final portion of the spinneret and the portion of the spinline up to $z = 5d_o$. The calculations were carried out by imposing a dimensionless force B at $z = 5d_o$ as the downstream boundary condition; the upstream flow was taken to be fully developed in the spinneret at $z = -2d_o$. The computed draw ratios for the Maxwell fluid are approaching the limiting values of 81 and 41 defined by the thin-filament analysis for $De = 1/80$ and $1/40$, respectively, with increasing force (increasing B), whereas the PTT fluid, which does not have a limiting value, shows a gradual increase. The values of De shown are the largest for which convergent solutions could be obtained. Contours of constant values of the velocity are shown in Figure 10.4 for the Maxwell fluid for increasing values of the dimensionless force B. There is flow rearrangement near the spinneret exit, and a uniform velocity profile is obtained on the spinline within one spinneret diameter downstream of the exit. A small degree of extrudate swell can be observed for $B = 0.5$, but the swell appears to be overcome by the tensile stress for $B = 1$. The ratio $-\tau_{rr}/\tau_{zz}$ at $z = 0$ was less than 0.2 for all B greater

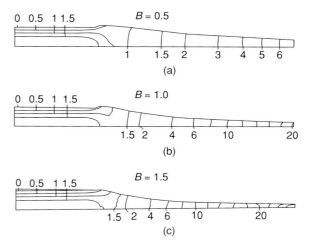

Figure 10.4. Contours of constant velocity for a Maxwell fluid with dimensionless forces B = (a) 0.5, (b) 1.0, and (c) 1.5 at z = $5d_o$. Reprinted with permission from Keunings et al., *Ind. Eng. Chem. Fundam.*, **22**, 347 (1983). Copyright American Chemical Society.

than 1.25, supporting the usual practice in thin filament calculations of taking this ratio to be zero.

Figure 10.5 shows a simulation of the poly(ethylene terephthalate) pilot plant experiments of George that were discussed in Section 7.4.4. PET is a difficult material to work with, and it has very small G' and normal stresses. No uniform uniaxial extensional data are available. As noted in Chapter 7, Gregory has published extensive data on the zero-shear viscosity and relaxation time of PET, where the relaxation time is defined as the reciprocal of the shear rate at which non-Newtonian

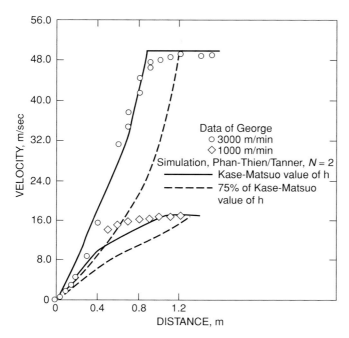

Figure 10.5. Simulation of George's PET pilot plant spinning experiments using the PTT model with $N = 2$. Reprinted from Gagon and Denn, *Polym. Eng. Sci.*, **21**, 844 (1981).

effects first appear, as follows:

$$\eta_o = 1.13 \times 10^{-14} M_w^{3.5} \exp(-11.98 + 6800/T) \text{ in Pa s.}$$
$$\lambda = 1.17 \times 10^{-9} M_w^{3.5} \exp(-11.98 + 6800/T) \text{ in s.}$$

Gagon and Denn estimated the linear viscoelastic parameters for two modes from Gregory's viscosity and relaxation time data by choosing a "wedge" spectrum, in which $G_i = \eta_o/N\lambda_i$, where N is the number of relaxation modes. They arbitrarily chose $\lambda_2/\lambda_1 = 5$. The parameter ξ was taken to be zero to reflect the shear insensitivity of the viscosity of PET, while ε was taken to be 0.015. The results were insensitive to ε in the range from 0 to 0.015, so the simulation was in effect carried out for a two-mode Maxwell model. The results depended on the choice of the coefficient for the air drag coefficient, but the variation was small in the range 0.37 to 0.6.

These experiments were simulated in Figure 7.5 using a Newtonian fluid model, where it was shown that the fit to the data was improved somewhat if the value of the heat transfer coefficient given by the Kase-Matsuo correlation was reduced by 25%. The same comparison is shown in Figure 10.5 for the PTT model. The fit to the data is now better using the original heat transfer correlation, and the quality of the fit is about the same as for the Newtonian fluid simulation. Devereux and Denn obtained essentially the same result by fitting the PTT model with a very different set of parameters, $\xi = 0.15$ and $\varepsilon = 0.2$, using a different spectrum shape with $\lambda_2/\lambda_1 = 56$.

These calculations point to a profound issue in the simulation of spinning for relatively inelastic melts like PET. Variations in the predicted results because of the uncertainty in the heat transfer coefficient are of the same magnitude as variations because of the inclusion or neglect of viscoelasticity. This is unimportant for spinning speeds below 4,000 m/min, where either way of modeling will predict the correct trends, and Newtonian models have been used very effectively in process analysis for PET spinning. It could become very important at significantly higher takeup speeds, however, where the stress development and the initiation of stress-induced crystallization could be quite different for the viscoelastic and Newtonian models. The viscoelastic model is clearly required for elastic polymers like polypropylene.

The major problem with the spinline simulations shown here, and with others like them, is the naïve way in which the solidification process is handled. It is simply assumed that the velocity no longer changes after a predetermined temperature has been reached. This approach has proved to be adequate for modeling PET at relatively low spinning speeds, where there is little crystallization and the solidification occurs at the glass transition temperature. It is less likely to be adequate for spinning nylon, which crystallizes rapidly. It has been shown to be inadequate for high-speed spinning of PET, where stress-induced crystallization occurs rapidly at a high temperature, but glass formation is also important. McHugh and co-workers have addressed this problem by adding a kinetic equation to the model for the formation

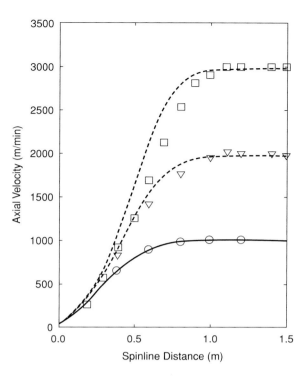

Figure 10.6. Simulation of George's PET pilot plant spinning experiments using a single-mode Giesekus fluid in a spinline model that incorporates crystallization. Reprinted with permission from Shrinkhande et al., *J. Appl. Polym. Sci.*, **100**, 3240 (2006). Copyright John Wiley & Sons, Inc.

and orientation of crystals under stress, together with a mixing rule for the stress in a biphasic system. The melt rheology is described either by a single-mode Giesekus model or by the tube-derived XPP model. Their crystallization model does a good job of matching George's PET data, as shown in Figure 10.6, as well as industrial nylon spinning data. What is particularly impressive, however, is the ability of the coupled model to predict the diameter profile in the high-speed spinning of PET, where the rapid onset of stress-induced crystallization causes a nearly discontinuous neck that cannot be described by the simple equations used here.

10.4 Film Blowing

The blown film process was briefly described in Section 1.2.6. The process is shown schematically in Figure 10.7. There are many similarities between the blown film and the fiber spinline because of the free surface and the very small transverse dimension relative to the distance between melt extrusion and solidification, and "thin sheet" equations analogous to the thin filament equations are typically used, although the hoop stress must now be taken into account. The equations for a Newtonian fluid were first published by Pearson and Petrie in 1970, and their approach has been used by nearly all investigators since. There are two steady-state momentum equations because variations in both thickness and width in the stretching direction are important. The mechanics of the solid region above the ill-defined freezeline are

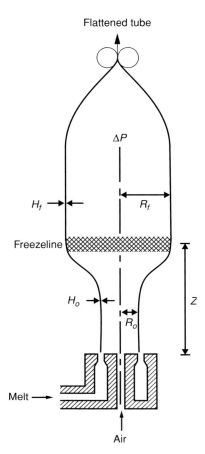

Figure 10.7. Schematic of the blown film process.

important, and details of the heat transfer, which may include radiation, are less well developed for film blowing than for fiber spinning. The air flow from an air ring at the point of exit from the annular die is very important regarding both heat transfer and shape and stabilization of the polymer bubble, but the air ring is rarely considered in simulations. The blown film process can be very sensitive to operate, and different bubble shapes, especially in the neck region, are sometimes observed under the same operating conditions.

There are two operating parameters, the dimensionless pressure B and the dimensionless axial takeup stress Tz, defined, respectively, as

$$B = \frac{R_0^2 \Delta P}{2\eta V_0 H_0}, \quad T_Z = \frac{F}{2\pi \eta H_0 V_0}.$$

ΔP is the pressure difference between the inside and outside of the bubble, and F is the takeup force. V_0 is the average melt velocity at the exit of the annular die, and the other geometrical variables are defined in the figure. The viscosity in the dimensionless groups is taken at a reference temperature, usually the feed. The thin sheet equations for the evolution of the dimensionless radius, r, and the dimensionless

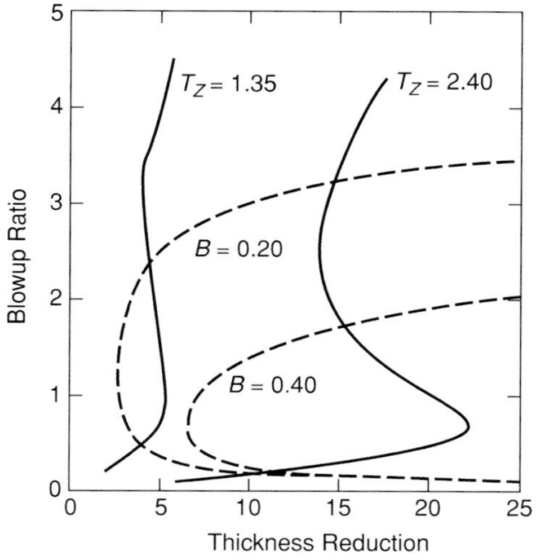

Figure 10.8. Operating parameter space for isothermal film blowing of a Newtonian fluid, $Z/R_0 = 5$. Reprinted from Cain and Denn, *Polym. Eng. Sci.*, **28**, 1528 (1988).

thickness, w, both normalized with respect to their initial values, are as follows for a Newtonian fluid:

$$2r^2[T - B(BUR)^2 + Br^2]r'' = 6r' + r(1 + r'^2)[T - B(BUR)^2 - 3Br^2], \qquad (10.7a)$$

$$\frac{w'}{w} = -\frac{r'}{2r} - \frac{(1 + r'^2)[T - B(BUR)^2 + Br^2]}{4}. \qquad (10.7b)$$

The *blowup ratio*, BUR, is the ratio of the final bubble radius to the initial radius and is an analog of the draw ratio in spinning. Primes denote differentiation with respect to the dimensionless axial length, which is normalized with respect to R_0. The differential equations are more complex for viscoelastic liquids, and the stress equations must be solved in parallel with the momentum equations. $r = w = 1$ at the exit from the die, and $r' = 0$ at the freezeline. The last boundary condition assumes that the viscosity becomes infinite at the freezeline and that there is no further deformation. (This condition is approximate at best and need not be used when a solidification model in which the solid phase evolves and locks in structure is employed.) Heat transfer is very important, although it has usually been handled with rather simplistic assumptions about the heat transfer coefficient.

The boundary value problem is usually solved using a "shooting" method, in which the third initial condition is assumed at the origin and adjusted with each iteration until the condition at the freezeline is satisfied. The system is very sensitive, and use of the shooting method precluded examination of the full parameter space for a long time after the initial work of Pearson and Petrie. The calculations shown here were carried out using an efficient finite difference algorithm designed for boundary value problems known as the *band method*. The operating parameter space for isothermal film blowing of a Newtonian fluid is shown in Figure 10.8 for a fixed distance to the freezeline. There are two possible operating states for a fixed dimensionless pressure difference (B) and fixed dimensionless takeup force (T_Z), one with a blowup ratio greater than unity and one in which the radius

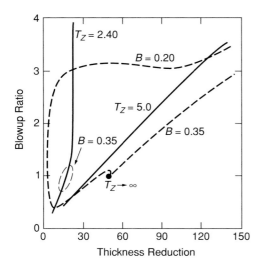

Figure 10.9. Operating parameter space for isothermal film blowing of a Maxwell fluid, $Z/R_0 = 5, \lambda V_0/R_0 = 0.1$. Reprinted from Cain and Denn, *Polym. Eng. Sci.*, **28**, 1528 (1988).

is drawn down. The shapes of the contours are changed somewhat by including heat transfer, but the qualitative results are the same. The shapes of the contours change for a Maxwell fluid, as shown in Figure 10.9; the constant pressure contours separate into two branches for B in excess of 0.23 for the parameters shown here. All contours converge to the point for which the blowup ratio is 1.0 and the thickness reduction ratio is $1 + Z/\lambda V_0$ as T_z grows without bound and the velocity profile becomes linear; this is the analog of the "unattainable" limiting behavior observed in spinning for the Maxwell fluid, as illustrated in Figure 10.2. The singular point for $T_Z \to \infty$ moves off to infinite thickness reduction and $BUR = 1$ if the Maxwell fluid is replaced by a fluid with bounded extensional stresses.

Multiple solutions with different bubble shapes and different freeze points can be found for the same values of B and T_Z when heat transfer is included with the Maxwell fluid. Figure 10.10 shows three different solutions obtained for a Maxwell

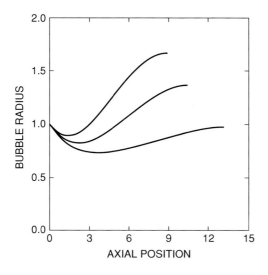

Figure 10.10. Multiple shape profiles for a Maxwell fluid with $\lambda V_0/R_0 = 0.335$ at $B = 0.061$ and $T_Z = 1.19$. Reprinted from Cain and Denn, *Polym. Eng. Sci.*, **28**, 1528 (1988).

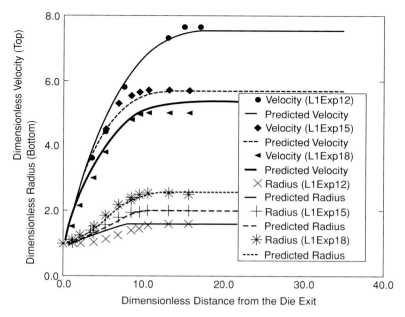

Figure 10.11. Comparison of model calculations of bubble radius and film velocity with blown film data for a low-density polyethylene, 3.84-cm radius × 0.8-mm thickness die, 4.1-kg/hr throughput. Data of Tas. Experiment 12: inflation pressure = 118 Pa, takeup force = 7.6 N. Experiment 15: 108 Pa, 7.7 N. Experiment 18: 95 Pa, 6.8 N. Reprinted with permission from Muslet and Kamal, *J. Rheol.*, **48**, 525 (2004).

fluid with $\lambda V_0 / R_0 = 0.335$ at $B = 0.0061$ and $T_z = 1.19$ using the thermal properties of polystyrene and a constant heat transfer coefficient. The parameters correspond to an experiment on blowing a polystyrene film by Gupta and Metzner, and the shortest profile was observed for experiments with the polystyrene.

Several investigators have modeled the blown film process with attention to the mechanics associated with crystallization and the elastic deformability of the material above the freezeline. Figure 10.11 shows predicted radius and velocity profiles from a simulation by Muslet and Kamal for an eight-mode Phan-Thien/Tanner viscoelastic fluid with parameters matched to the linear viscoelastic and steady shear rheology of the low-density polyethylene used in the film experiments; $\xi = 0.15$ and $\varepsilon = 0.05$ for all modes. The model includes allowance for the temperature variation across the thickness, and it contains a crystallization model, but it still utilizes the final condition $r' = 0$. The heat transfer coefficient is not in a form that scales with the usual dimensionless groups. The agreement between the model predictions and experimental data for three inflation pressures is quite good, and similar good agreement was obtained with other experiments. Predictions of the average degree of crystallinity were also in good agreement with experimental data. Henrichsen and McHugh have analyzed film blowing using their two-phase formulation to account for crystallization, which does not require the condition $r' = 0$ to define the freeze point. They showed very good agreement with a different data set on both

Table 10.1. *Phan-Thien/Tanner model parameters used by Verbeeten and co-workers for DSM Stamylan LD 2008 XC43 LDPE ($M_w = 1.55 \times 10^5$, $M_w/M_n = 11.9$) at $T = 170\,°C$*

Mode	λ_i (s)	G_i (Pa)	ε_i	ξ_i
1	3.89×10^{-3}	7.20×10^4	0.30	0.08
2	5.14×10^{-2}	1.58×10^4	0.20	0.08
3	5.03×10^{-1}	3.33×10^3	0.02	0.08
4	4.59	3.01×10^2	0.02	0.08

low-density and linear low-density polyethylene films. Their heat transfer coefficient was also empirical and not easily scalable.

10.5 Converging Flow

Flow through a converging section is a common feature of polymer melt processes. The flow along the centerline is purely extensional, so we may expect large stresses to develop. In addition, it is well known that recirculating vortices develop in corners, even for Newtonian fluids. The vortices typically grow with increasing flow rate (increasing Wi) for branched polymers, becoming much larger than the Newtonian fluid vortex, but not for linear polymers. Flow through a contraction has been the most widely studied application of computational fluid dynamics for viscoelastic liquids because of the importance of the problem and because of the challenge presented by the corner singularity at the contraction entrance and the apparent dependence of the vortex structure on chain architecture. The example we use here is by Verbeeten and co-workers from the polymer processing group at the Technical University of Eindhoven.

The polymer is a branched polyethylene melt with $M_w = 1.55 \times 10^5$ and $M_w/M_n = 11.9$, flowing at $170\,°C$ in a 3.3:1 planar contraction. The linear viscoelastic properties and the nonlinear parameters for a four-mode PTT equation are shown in Table 10.1. Different values of ε were used for each mode, but with a constant value of $\xi = 0.08$. These parameters provide a reasonable fit to the transient and steady-state shear and extensional data, although the nonlinear parameters for the two longest relaxation times cause small oscillations in startup of simple shear that are not observed experimentally; using parameters that eliminate the shear oscillations causes the calculated extensional stresses to be too low, and the contraction flow results are sensitive to the extensional stresses. The mean relaxation time was 1.74 s, the average velocity in the downstream channel was 7.47 mm/s, and the downstream channel half-width (the characteristic length) was 0.775 mm. The Weissenberg number based on downstream channel properties was therefore 16.8.

Measured and computed isochromatic fringe patterns, which are contours of the characteristic stress $(N_1^2 + 4\tau_{xy}^2)^{1/2}$, are shown in Figure 10.12. The experimental values were obtained from birefringence measurements. The experimental and computed fringe patterns are in good agreement, including the large corner vortices,

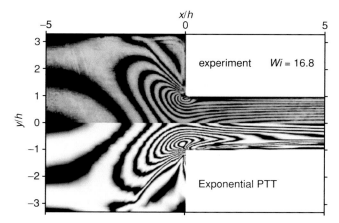

Figure 10.12. Experimental and calculated isochromatic fringe patterns for planar converging flow of a low-density polyethylene melt, $Wi = 16.8$, $T = 170\,^{\circ}$C. Reprinted with permission from Verbeeten et al., *J. Non-Newtonian Fluid Mech.*, **117**, 73 (2004). Copyright Elsevier.

the concentration of stresses near the reentrant corner, and the "butterfly" pattern. The measured and computed characteristic stresses along the centerline and across the contraction plane are shown in Figure 10.13, which also includes results from a version of the pom-pom model (XPP). There is little difference between the stress predictions of the two models, although the PTT model with the selected parameters gives a better overall fit to the isochromatic fringes.

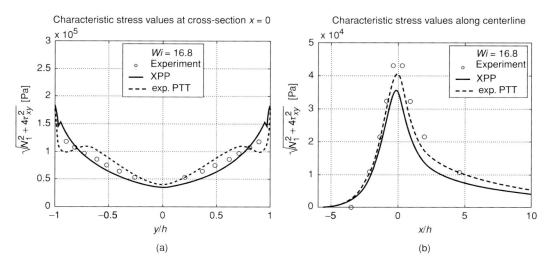

Figure 10.13. Experimental and calculated characteristic stresses for planar converging flow of a low-density polyethylene melt, $Wi = 16.8$, $T = 170\,^{\circ}$C: (a) along the centerline; (b) at the plane of convergence. Reprinted with permission from Verbeeten et al., *J. Non-Newtonian Fluid Mech.*, **117**, 73 (2004). Copyright Elsevier.

Table 10.2. *Linear viscoelastic parameters used by Baaijens and co-workers for DSK Stamylan LD 2008 XC43 LDPE* ($M_w = 1.55 \times 10^5$, $M_w/M_n = 11.9$) *at* $T = 190\,°C$

Mode	λ_i (s)	G_i (Pa)
1	3.16×10^{-5}	2.69×10^5
2	1.00×10^{-3}	1.37×10^5
3	3.16×10^{-2}	2.09×10^4
4	1.00	1.65×10^3

10.6 Flow Past an Obstruction

As a final example of viscoelastic simulation of melt flow in a complex geometry, we consider flow of the same branched polyethylene, now at $190\,°C$, in a planar channel in which a cylinder has been placed asymmetrically. The example is by J. P. W. Baaijens, also from the polymer processing group at Eindhoven. The cylinder has a diameter of 2.5 mm, which is the same as the half-width of the channel. The cylinder center is placed 1.75 mm from one surface, so the gaps between the cylinder and the two walls are 0.5 and 2 mm. A geometry of this type might arise in a mold, for example. This is a complex flow. There is a high level of shear in the gaps between the cylinder and the wall, as well as flow development both approaching and leaving the gaps. There are also fore and aft stagnation points on the cylinder; the flow near the upstream stagnation point is approximately compressive, while the flow near the downstream stagnation point is approximately elongational, so large stresses are anticipated.

The mean velocity in the upstream and downstream channel was 8.53 mm/s. The linear viscoelastic parameters at $190\,°C$ are shown in Table 10.2; the mean relaxation time is 0.68 s. The Weissenberg number in the region of fully developed flow upstream and downstream of the cylinder was therefore around 2. The flow should distribute between the two gaps roughly as the gap ratio cubed (cf. Equation 5.8 for Newtonian fluids), so the average velocity in the 0.5 mm gap was approximately 1.3 mm/s, while the average velocity in the 2.0 mm gap was approximately 21 mm/s. The characteristic Weissenberg numbers for the small and large gaps were then, respectively, 0.8 and 14. The mean residence time in each gap was on the order of the mean velocity divided by the cylinder diameter, so characteristic Deborah numbers for the small and large gaps were, respectively, 0.3 and 5.3, indicating the likelihood of strong elastic effects.

The experimental isochronal fringes are shown in Figure 10.14. Baaijens carried out finite element calculations for PTT fluids with $\xi = 0.1$, $\varepsilon = 0.1$ and $\xi = 0.2$, $\varepsilon = 0$, using the same values of ξ and ε for all modes, as well as for a Giesekus fluid. Both sets of PTT parameters were adequate for the shear data, and no elongational data were available at $190\,°C$. The computed results for $\xi = 0.1$, $\varepsilon = 0.1$ are superimposed on the experiments in Figure 10.14. Convergence could not be obtained

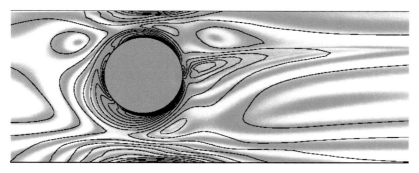

Figure 10.14. Experimental and calculated isochromatic fringe patterns for planar channel flow of a branched polyethylene melt past a cylinder, $T = 190\,^{\circ}\mathrm{C}$: PTT fluid with four modes, $\xi = 0.1$ and $\varepsilon = 0.1$. Reprinted with permission from F. P. T. Baaijens et al., *J. Non-Newtonian Fluid Mech.*, **68**, 173 (1997). Copyright Elsevier.

at a higher Weissenberg number with either the second set of PTT parameters or the Giesekus model. The computations are in qualitative agreement with the experiments and show the development of an asymmetric wake in which the stresses are very high downstream of the aft stagnation point. The experimental wake extends farther in the downstream direction than the calculated wake. The fringes appear to be more densely packed in the lower upstream region close to the cylinder in the experiments than in the calculations, indicating that the experimental stress gradients were stronger than the computed gradients.

10.7 Secondary Flows

Viscoelasticity can cause changes in flow fields that are unexpected from our experience with Newtonian liquids. It can be shown in complete generality, for example, that rectilinear flow (i.e., flow streamlines that are parallel to the wall) is possible in a closed channel of constant noncircular cross section only if the second normal stress difference is zero. There *must* be a transverse flow if the second normal stress difference is nonzero.

Figure 10.15 shows an elegant study by Dooley in which a special feedblock was used to create continuous layers of colored and uncolored high-impact polystyrene (Dow Styron 484) with identical properties at the entrance to a 61-cm-long channel with a square cross section that is 0.9525 cm on a side. The thirteen concentric rings in Figure 10.15a evolved to the complex pattern in Figure 10.15b because of the transverse flow. The computed secondary flow using a finite element code is shown in Figure 10.15c. The published computations, which are two-dimensional since the flow is the same at all cross sections, were done using a Giesekus model, but a Phan-Thien/Tanner model was reported to have given the same result.

The motivation for this study was interface movement in the coextrusion of multilayer films. Figure 10.16 shows the computed and experimental deformation of the interface for flow of a low-density polyethylene (Dow 641I) through the same 61-cm channel. Here, only two layers were used. A three-dimensional finite element

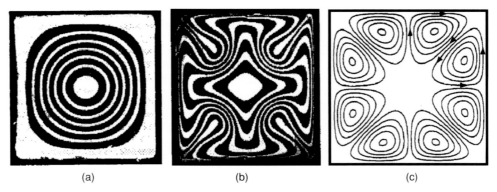

Figure 10.15. Flow of a high-impact polystyrene resin (Dow Styron 484) through a 61-cm channel with a square cross section that is 0.9525 cm on a side. A special feedblock brings in concentric layers, which are colored for contrast. (a) At the channel entrance. (b) At the channel exit. (c) Computed secondary flow. Reprinted with permission from J. Dooley, *Viscoelastic Flow Effects in Multilayer Polymer Coextrusion*, Ph.D. dissertation, Technical University of Eindhoven, the Netherlands, 2002.

computation was done using a five-mode Giesekus fluid; a two-dimensional calculation could have been used to track the interface movement, but the 3-D code was being tested. The agreement between the computed and experimental interface evolution is quite good, despite some quantitative differences.

The example is an impressive demonstration both of a physical phenomenon that is driven by nonlinear viscoelasticity and of the use of numerical simulation. It is probably not the dominant mechanism for interface movement in multilayer systems, however, where viscosity and normal stress jumps across the interface are more important than the rather weak effect of the nonzero second normal stress difference.

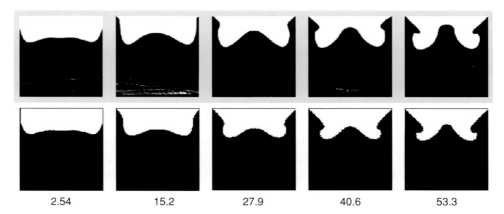

Figure 10.16. Flow of a low-density polyethylene (Dow 641I) through a 61-cm channel with a square cross section that is 0.9525 cm on a side. The two layers are identical except for the color to define the interface. Cross sections were cut at the positions shown. Top: numerical simulation; Bottom: experiment. Reprinted with permission from Debbaut and Dooley, *J. Rheol.*, **43**, 1525 (1999).

10.8 Concluding Remarks

The flows discussed in this chapter are good examples of how numerical simulation can be used to predict streamlines and stress levels for viscoelastic polymer melts in complex geometries. Qualitative agreement is very good, and quantitative agreement is sufficient to make informed engineering judgments. In keeping with the general introductory level of this text, we have omitted any discussion of strictly numerical issues, other than to note that convergence can be difficult to achieve in flows with strong singularities at a boundary or a stagnation point, where large stress gradients that can propagate through the flow must be resolved. The more realistic the constitutive equation, the better the convergence seems to be, probably because realistic constitutive equations exhibit smaller stress gradients at singularities than, say, the Maxwell model. Improvement between simulations and experiments can be expected with the growing use of constitutive equations that incorporate molecular information, such as the pom-pom model. The article by Lee and co-workers listed in the Bibliographical Notes is a good example of the application of the differential form of the pom-pom model to flow through a contraction followed by an expansion.

Some important issues have not been addressed at all here. One is the question of the correct formulation of the wall boundary condition in a region of very high stress, where chain disentanglement or adhesive failure may occur in the wall region and the conventional no-slip condition may fail. This topic is addressed in Chapter 12. Another is the modeling of flows of polymer melts that have rigid backbones or are highly filled with particulates and cannot be described by constitutive equations of the general class employed here. This topic, which is still in a rather undeveloped state, is addressed briefly in Chapter 13. Finally, computation for mixing and dispersion is touched on in Chapter 14.

BIBLIOGRAPHICAL NOTES

There is a good but dated tutorial review, including a discussion of the high Weissenberg number problem, in

Keunings, R. "Simulation of Viscoelastic Fluid Flow," in C. L. Tucker, III, Ed., *Fundamentals of Computer Modeling for Polymer Processing*, Hanser, Munich, 1989, pp. 403ff.

There has been considerable progress since Keuning's review was completed in 1987. A good review of technical issues can be found in

Baaijens, F. P. T., *J. Non-Newtonian Fluid Mech.*, **79**, 361 (1998).

The subject is addressed broadly and in depth in a relatively recent text,

Owens, R. G., and T. N. Phillips, *Computational Rheology*, Imperial College Press, London, 2002.

Methods for integral constitutive equations are reviewed in

Keunings, R., "Finite Element Methods for Integral Viscoelastic Fluids," in K. Walters, Ed., *Annual Rheology Reviews*, British Society of Rheology, London, 2002.

See also a review of consortium activities by twenty-two authors,

Agassant, J. F. et al., *Int. Polym. Proc.*, **XVII**, 3 (2002).

The field continues to advance, with international workshops held every two years, and recent issues of the *Journal of Non-Newtonian Fluid Mechanics*, the *Journal of Rheology*, and *Rheologica Acta* should be consulted, with the first of the three serving as the primary venue. Applications articles can be found in *Polymer Engineering and Science* and *International Polymer Processing*.

There is an extensive discussion of extrudate swell in

Tanner, R. I., *Engineering Rheology,* 2nd ed., Oxford, New York, 2000, pp. 418ff.

Annular extrudate swell is important for parison formation in blow molding. Finite element calculations and comparison to experiment, together with calculations and measurements of blown bottle thickness, are in

Yousefi, A.-M., P. Collins, S. Chang, and R. W. DiRaddo, *Polym. Eng. Sci.*, **47**, 1 (2007).

There is more detail of the calculations in a numerical study of parison inflation in

Debbaut, B., O. Homerin, and N. Jivraj, *Polym. Eng. Sci.*, **39**, 1812 (1999).

Spinline modeling through 1983 is reviewed in the following, which includes the results from Gagon and Denn shown here:

Denn, M. M., "Fibre Spinning," in J. R. A. Pearson and S. M. Richardson, Eds., *Computational Analysis of Polymer Processing*, Applied Science, London, 1983, pp. 179ff.

The alternate parameters for PET spinline simulation, as well as parameters for polypropylene, are developed in

Devereux, B. M., and M. M. Denn, *Ind. Eng. Chem. Res.*, **33**, 2384 (1994).

The focus of this latter work is spinline dynamics, which we address in the next chapter. The comparisons to George's PET data are in an appendix to Devereux and Denn that is available as supplementary material. The work of McHugh and co-workers incorporating crystallization kinetics and the contribution of the crystalline phase to the stress can be found in

Doufas, A. K., and A. J. McHugh, *J. Rheol.*, **45**, 403 and 855 (2001).
Shrinkhande, P., W. H. Kohler, and A. J. McHugh, *J. Appl. Polym. Sci.*, **100**, 3240 (2006).
Kohler, W. H., and A. J. McHugh, *J. Rheol.*, **51**, 721 (2007).

The two 2001 *Journal of Rheology* articles contain PET simulations and references to earlier work. The first article employs the thin filament equations, while the second incorporates the 2-D energy equation. The later article by Shrinkhande et al. modifies the model to remove a discontinuity and includes the PET simulations shown here, while the last article includes the 2-D energy equation to account for crystallization gradients and radial birefringence profiles. Kannan and co-workers introduced a model that includes the dynamics of the glass transition and a mixing rule into the thin filament equations, and they have also compared the results to George's PET data; see

Kannan, K., I. J. Rao, and K. R. Rajagopal, *J. Rheol.*, **46**, 977 (2002).

Simulations of nylon and PET spinning using a one-phase crystallization model and a PTT equation are in

Shin, D. M., J. S. Lee, H. W. Jung, and J. C. Hyun, *Rheol. Acta*, **45**, 575 (2006).

The mechanics of film blowing are reviewed in

Pearson, J. R. A., *Mechanics of Polymer Processing*, Elsevier Applied Science Publishers, London, 1985;

in the book by Tanner, cited above; and in a chapter by Petrie:

Petrie, C. J. S., "Film Blowing, Blow Moulding, and Thermoforming," in J. R. A. Pearson and S. M. Richardson, Eds., *Computational Analysis of Polymer Processing*, Applied Science Publishers, London, 1983, pp. 217ff.

As the title indicates, Petrie's chapter also addresses work through 1983 on blow molding and thermoforming. The Newtonian and Maxwell fluid results shown here are from the following, which also includes results for still another Maxwell-type equation with bounded extensional stresses (the *Marrucci* model):

Cain, J. J., and M. M. Denn, *Polym. Eng. Sci.*, **28**, 1527 (1988).

There is a transcription error in formulating the transient equations in this work that makes the dynamical results uncertain, but the transient does not concern us here. Simulations using the PTT model that show multiplicities, together with experimental multiplicity data, are in

Shin, D. M., J. S. Lee, H. W. Jung, and J. C. Hyun, *J. Rheol.*, **51**, 605 (2007).

There is an extensive treatment of the parameter space for models including the PTT fluid in

Housiadis, K. D., G. Klidis, and J. Tsamopoulos, *J. Non-Newtonian Fluid Mech.*, **141**, 193 (2007).

References to earlier literature may be found in all of the articles cited above. There is a very good overview of blown film modeling in a French doctoral dissertation,

André, J.-M., *Modelisation Thermomécanique et Structurale du Soufflage de Gaine de Poly-éthylènes*, Thèse, Ecole Nationale Supérieure de Physique de Grenoble, 1999,

and there are two brief publications based on this work:

André, J.-M., Y. Demay, and J.-F. Agassant, *C. R. Acad. Sci. Paris*, **325**, **Ser. II**, 621 (1997).
André, J.-M., J.-F. Agassant, Y. Demay, J.-M. Haudin, and B. Monasse, *Int. J. Forming Proc.*, **1**, 187 (1998).

The first of these articles, which is in French, deals with multiplicities. The simulation of the low-density polyethylene data using the Phan-Thien/Tanner viscoelastic liquid with crystallization is from

Muslet, I. A., and M. R. Kamal, *J. Rheol.*, **48**, 525 (2004).

A film blowing application of the crystallization model first developed for spinning in the articles by Doufas and McHugh, cited above, with comparisons to spinline data for low-density and linear low-density polyethylene, is in

Henrichsen, L. K., and A. J. McHugh, *Int. Polym. Proc.*, **XXII**, 179 (2007).

There were many qualitative studies of entry flows of polymer melts in the 1960s and 1970s. These are reviewed in

White, J. L., *Appl. Polym. Symp.*, **20**, 155 (1973).
Petrie, C. J. S., and M. M. Denn, *AIChE J.*, **22**, 209 (1976).

There are excellent photographs of entry flows of viscoelastic liquids, mostly polymer solutions, as well as flows in other complex geometries in

Boger, D. V., and K. Walters, *Rheological Phenomena in Focus*, Elsevier, Amsterdam, 1993.

Computational results and references can be found in the 1998 article by Baaijens, cited above, and in a chapter on "Benchmark Problems" in the text by Owens and Phillips. The example of entry flow used here is from

Verbeeten, W. M. H., G. W. M. Peters, and F. P. T. Baaijens, *J. Non-Newtonian Fluid Mech.*, **117**, 73 (2004).

Computations to evaluate the use of entry flows for deducing extensional viscosities can be found in

Mitsoulis, E., S. G. Hatzikiriakos, K. Christodoulou, and D. Vlassopoulos, *Rheol. Acta*, **37**, 438 (1998).
Rajagopalan, D., *Rheol. Acta*, **39**, 138 (2000).

See also an exchange of letters between the authors of these two articles in *Rheologica Acta*, **40**, 401 and 504 (2001).

Flow past a cylinder in a channel is another of the benchmark problems discussed in Owens and Phillips. The experiments and calculations shown here are from

Baaijens, F. P. T., S. H. A. Selen, H. P. W. Baaijens, G. W. M. Peters, and H. E. H. Meier, *J. Non-Newtonian Fluid Mech.*, **68**, 173 (1997).

There are small differences between the Weissenberg and Deborah numbers quoted in the article and the numbers used here, but the differences are unimportant in terms of the overall conclusions. Computations of nonisothermal flow for a symmetric cylinder, without experiments, are described in

Baaijens, F. P. T., and G. W. M. Peters, *J. Non-Newtonian Fluid Mech.*, **68**, 205 (1997).

The study of transverse flows is from

Dooley, J., *Viscoelastic Flow Effects in Multilayer Polymer Coextrusion*, Ph.D. dissertation, Technical University of Eindhoven, the Netherlands, 2002.

Some of the results are contained in two articles,

Debbaut, B., T. Avalosse, J. Dooley, and K. Hughes, *J. Non-Newtonian Fluid Mech.*, **69**, 255 (1997).
Debbaut, B., and J. Dooley, *J. Rheol.*, **43**, 1525 (1999).

Viscoelastic effects in some free-surface flows have been analyzed using very elementary constitutive models. For the dynamics of surface tension-driven breakup of a viscoelastic filament, see

Bousfield, D. W., R. Keunings, G. Marrucci, and M. M. Denn, *J. Non-Newtonian Fluid Mech.*, **21**, 79 (1986).

Transient elongational recovery of a polymer melt following stretching is in

Langouche, F., and B. Debbaut, *Rheol. Acta*, **38**, 48 (1999).

Squeeze flow of a viscoelastic liquid, including the movement of the free edge and non-isothermal effects, has been studied in

Debbaut, B., *J. Non-Newtonian Fluid Mech.*, **98**, 15 (2001).

A good example of the use of the differential pom-pom model to simulate a complex flow is

Lee, K., M. R. Mackley, T. C. B. McLeish, T. M. Nicholson, and O. G. Harlen, *J. Rheol.*, **45**, 1261 (2001).

See also the reviews by Baaijens and Agassant and co-workers, the 2004 article by Verbeeten and co-workers cited above, and

Van Os, R. G. M., and T. N. Phillips, *J. Non-Newtonian Fluid Mech.*, **129**, 142 (2005).

11 Stability and Sensitivity

11.1 Introduction

Our discussion of continuous processes like extrusion and spinning has focused thus far on steady operation. The dynamical response of these processes is also an important processing consideration. The field of process dynamics has paid little attention to polymer processing, other than to apply classical control system methodology to implement temperature control loops. In particular, models of continuous processes have not been used extensively, and there is considerable scope for dynamical analyses to improve operation and control.

There are two fundamental issues in considering the dynamics of a process. One is operational *stability*: If we design a process to operate under given conditions, and the process moves away from the design conditions for any reason, will it ultimately return or will it move further away? The other is operational *sensitivity*: If the process is operating under the design conditions, and disturbances enter the system, will the disturbances attenuate or will they grow as they propagate through the process? These are different questions, although they are often treated as the same because, up to a point, they share a common mathematical framework.

The dynamics of melt spinning has received more attention than any other process, and it is the primary focus of this chapter, although other processes are also briefly addressed. Instabilities in rectilinear flow through an extrusion die are confounded by issues regarding boundary conditions at high stress levels, and they are addressed in the next chapter.

11.2 Draw Resonance

Draw resonance, also known as *melt resonance*, is a phenomenon that occurs under certain conditions on a melt spinline when the extrusion rate and the takeup speed are both constant. Figure 11.1 shows the diameter of a drawn filament of poly(ethylene terephthalate) as a function of distance along the filament under

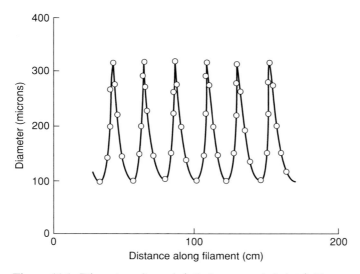

Figure 11.1. Diameter of a poly(ethylene terephthalate) filament as a function of distance along the filament, $L = 10$ cm, $D_R = 50$. Reprinted with permission after Ishihara and Kase, *J. Appl. Polym. Sci.*, **20**, 169 (1976). Copyright John Wiley & Sons, Inc.

constant spinning conditions at a draw ratio of about 50. The filament was relatively short in an attempt to minimize cooling in the melt zone, and it was then passed through a water bath to effect rapid solidification. Draw resonance occurs infrequently in commercial spinning, since it does seem to require a nearly isothermal environment in the draw zone, followed by rapid solidification, and this is not a typical commercial spinning configuration (although it is the configuration for the "gel spinning" of high-strength fibers from liquid crystalline solutions of poly(p-phenylene terephthalamide), sold commercially as Kevlar® and Twaron®, where the draw region is very much like that in melt spinning and mass transfer considerations arise only after contact with a quench bath). Draw resonance is a common problem in the mechanically similar two-dimensional process of extrusion coating, where the draw zone is short and solidification occurs when the melt strikes the moving sheet. Indeed, the phenomenon first came to widespread attention because of reports of periodic coating thickness variations that destroyed the optical quality of coated sheets.

One's intuition is to seek a mechanism for a resonant interaction (hence, the term "draw resonance") between some forcing disturbance and the viscoelastic time scale of the polymer melt. This approach is misplaced; experiments with a variety of polymers clearly show that the characteristic period for the diameter oscillations is uncorrelated from any characteristic relaxation time of the polymer, but the period does correlate quite well with the residence time on the spinline. In fact, draw resonance can occur even for a Newtonian fluid; Figure 11.2 shows periodic tension variations in the steady drawing of a filament of corn syrup, a Newtonian liquid, at a draw ratio of 26. (The takeup device was a roll from which the liquid

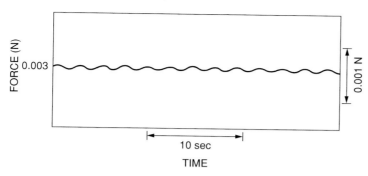

Figure 11.2. Force as a function of time for continuous drawing of a filament of a Newtonian corn syrup, $D_R = 26$. Reprinted from Chang and Denn, *J. Non-Newtonian Fluid Mech.*, **5**, 369 (1979).

corn syrup was scraped; hence, it was not possible to obtain data on diameter variations.)

11.3 Linear Stability Analysis of Melt Spinning

Draw resonance is a particularly good example of a processing instability because it has a clear signature and a sharp onset. Furthermore, the onset is amenable to rigorous analysis because of the simplicity of the thin filament spinning equations developed in Chapter 7. The approach we use here is *linear stability theory*, which asks a very specific question:

> *Suppose a system is designed to operate at steady state, and for some reason it is perturbed infinitesimally away from the steady-state conditions. Will the system return to the steady state in time, or will it move even further away?*

The steady state is unconditionally unstable and cannot be maintained if the answer is that the system will continue to move away following any infinitesimal disturbance. The conclusion is less definitive if the system is found to return to the steady state following an infinitesimal disturbance, since we have no way of knowing what will happen if the disturbance is finite. We never learn what ultimately happens to an unstable system using this approach, since we are enquiring only about the dynamics of infinitesimal disturbances, and a growing disturbance will cease to be infinitesimal.

The linear stability approach can be illustrated by a simple example. Consider the first-order ordinary differential equation

$$\frac{dv}{dt} = Kv(1 - v), \ K > 0. \tag{11.1}$$

The steady state, v_s, is the solution to Equation 11.1 when $dv/dt = 0$:

$$Kv_s(1 - v_s) = 0, \tag{11.2}$$

which has two solutions, $v_s = 0$ and 1. We now change the frame of reference to the steady state by defining a new variable $u(t)$ such that $v(t) = v_s + u(t)$; this change of variables permits us to focus on the perturbation. Equation 11.1 is then rewritten

$$\frac{du}{dt} = K(v_s + u)(1 - v_s - u) = Kv_s(1 - v_s) + Ku(1 - v_s) - Kv_s u + O(u^2), \quad (11.3)$$

where $O(u^2)$ denotes terms that go to zero at least as fast as u^2. The $O(u^2)$ terms arise naturally in this example because the nonlinearity is quadratic; in other cases it might be necessary to expand a nonlinear function in a Taylor series to obtain this form. The term $Kv_s(1 - v_s)$ vanishes by virtue of Equation 11.2.

We now assume that $|u|$ is sufficiently small that terms of order u^2 are negligible relative to terms of order u. In that case we will neglect the $O(u^2)$ terms in Equation 11.3 and write

$$\frac{du}{dt} = Ku(1 - 2v_s). \quad (11.4)$$

The solution to Equation 11.4 is

$$u(t) = u(0) \exp[K(1 - 2v_s)t], \quad (11.5)$$

where $u(0)$ is an arbitrary (infinitesimal) disturbance at $t = 0$. The solution is a growing exponential for $v_s = 0$, so $u(t)$ grows in magnitude; hence, $v_s = 0$ is absolutely unstable, since $v(t)$ moves away from v_s following any arbitrarily small perturbation. The solution is a decaying exponential for $v_s = 1$, so this steady state is conditionally stable. (We know for this equation that the solution will approach $v = 1$ as $t \to \infty$ for any initial condition $v(0) > 0$, so in fact $v_s = 1$ is absolutely stable, but that conclusion does not follow from the linearized analysis that we have done here.)

We now turn to the transient thin filament equations for a spinline, as given in Appendix 7B. For simplicity, we will restrict ourselves to isothermal spinning of a Newtonian fluid in the absence of inertia, air drag, or gravity, in which case the relevant equations are

$$\text{conservation of mass:} \quad \frac{\partial A}{\partial t} = -\frac{\partial (Av)}{\partial z}, \quad (11.6a)$$

$$\text{conservation of momentum:} \quad 0 = -\frac{\partial (A\sigma)}{\partial z}, \quad (11.6b)$$

$$\text{constitutive equation:} \quad \sigma = 3\eta \frac{\partial v}{\partial z}. \quad (11.6c)$$

A is the cross-sectional area. We have dropped the convention of overbars to denote averaged quantities. σ is the axial stress, σ_{zz}. We can, of course, combine Equations 11.6b and c – indeed, we can combine the three equations into a single nonlinear integro-differential equation – but we retain this structure to facilitate comparison with the corresponding analysis for viscoelastic liquids, where the stress and momentum equations cannot be combined.

The steady-state solution to Equations 11.6a–c is given by Equations 7.45a–d:

$$A_s = A_o \exp\left(-z \ln D_R/L\right), \tag{11.7a}$$

$$v_s = v_o \exp\left(z \ln D_R/L\right), \tag{11.7b}$$

$$\sigma_s = \frac{3\eta v_o \ln D_R \exp\left(z \ln D_R/L\right)}{L}. \tag{11.7c}$$

Note that unlike the elementary example employed to illustrate linearization, the steady state on the spinline is a function of position. We now define new dependent variables in order to use the position-dependent steady state as the frame of reference:

$$A(z, t) = A_o\left[\exp\left(-z \ln D_R/L\right) + \Phi(z, t)\right], \tag{11.8a}$$

$$v(z, t) = v_o\left[\exp\left(z \ln D_R/L\right) + \Psi(z, t)\right], \tag{11.8b}$$

$$\sigma(z, t) = \frac{3\eta v_o}{L}\left[\ln D_R \exp\left(z \ln D_R/L\right) + \Pi(z, t)\right]. \tag{11.8c}$$

We also define new independent variables,

$$\zeta = z/L, \quad \theta = v_o t/L. \tag{11.9}$$

We introduce the change of variables into Equations 11.6a–c and neglect nonlinear terms in the perturbation variables Φ, Ψ, and Π. (The nonlinearities here are quadratic, but they will not be quadratic for the energy equation or for a viscoelastic constitutive equation like the PTT model.) We thus obtain the following linear equations:

$$\frac{\partial \Phi}{\partial \zeta} = -\exp\left(-\zeta \ln D_R\right)\frac{\partial \Phi}{\partial \theta} - \ln D_R \Phi + \exp\left(-2\zeta \ln D_R\right)\left[\Psi + \ln D_R \Psi\right], \tag{11.10a}$$

$$\frac{\partial \Pi}{\partial \zeta} = \ln D_R \Pi - \ln D_R \exp\left(2\zeta \ln D_R\right)\left[\frac{\partial \Phi}{\partial \zeta} - \ln D_R \Phi\right], \tag{11.10b}$$

$$\frac{\partial \Psi}{\partial \zeta} = \Pi. \tag{11.10c}$$

The velocity is specified at $\zeta = 0$ and $\zeta = 1$; thus, the velocity perturbation at these points must be zero. Similarly, the initial area is fixed, so the area perturbation must vanish at $\zeta = 0$. The boundary conditions for these linear first-order partial differential equations are therefore

$$\Psi(0, t) = \Psi(1, t) = \Phi(0, t) = 0. \tag{11.11}$$

Time enters these equations only through the derivative $\partial \Phi/\partial \theta$. The time dependence of Φ must therefore be exponential. Furthermore, Ψ and Π must have the same time dependence as Φ, since linear combinations of these terms sum to zero at all times. Hence, we expect solutions to be linear combinations (because of the linearity of the equations) of terms of the form $\exp(\Lambda t)$ multiplied by a function of ζ for each of the three variables. We therefore seek solutions of the form

$$\Phi(\zeta, t) = e^{\Lambda t}\phi(\zeta), \tag{11.12a}$$

$$\Pi(\zeta, t) = e^{\Lambda t}\varpi(\zeta), \tag{11.12b}$$

$$\Psi(\zeta, t) = e^{\Lambda t}\psi(\zeta). \tag{11.12c}$$

(ϖ is a lowercase form of the Greek letter pi.) Upon substitution into Equations 11.10a–c we then obtain a set of linear ordinary differential equations,

$$\frac{d\phi}{d\zeta} = -\Lambda \exp(-\zeta \ln D_R)\phi - \ln D_R\phi + \exp(-2\zeta \ln D_R)[\psi + \ln D_R\psi], \tag{11.13a}$$

$$\frac{d\varpi}{d\zeta} = \ln D_R\varpi - \ln D_R \exp(2\zeta \ln D_R)\left[\frac{\partial\phi}{\partial\zeta} - \ln D_R\phi\right], \tag{11.13b}$$

$$\frac{d\psi}{d\zeta} = \varpi, \tag{11.13c}$$

with boundary conditions

$$\psi(0) = \psi(1) = \phi(0) = 0. \tag{11.14}$$

This system of linear, homogeneous equations always admits the trivial solution $\psi = \phi = \varpi = 0$, but we are interested only in nonzero solutions. For the reader to whom this is a new concept, we refer to the analogous problem for a system of linear algebraic equations of the form $\mathbf{Au} = \Lambda\mathbf{u}$. This system always has the trivial solution $\mathbf{u} = \mathbf{0}$, but nontrivial solutions will exist when the determinant of the matrix $\mathbf{A} - \Lambda\mathbf{I}$ equals zero, where $\mathbf{I}$ is the identity matrix. The vanishing of the determinant leads to an algebraic equation for Λ that has as many roots as the order of the matrix $\mathbf{A}$. These roots are known as characteristic values, or *eigenvalues*. Similarly, Equations 11.13a–c with the boundary conditions Equation 11.14 will admit nonzero solutions for certain characteristic values Λ, and we seek these values. If all such characteristic values are negative, then all terms of the form of Equations 11.12a–c comprising the solution to the differential equations will be negative exponentials, and the perturbation will die out in time. On the other hand, if *any one* such characteristic value Λ is positive, then that term will grow exponentially and the solution will diverge from the steady state for any disturbance, no matter how small. The characteristic values are in fact complex numbers, so these comments apply to the real parts; the exponential of an imaginary number is oscillatory.

A variety of solution methods can be applied to solve the eigenvalue problem. The most straightforward, which generalizes to the more interesting problem that includes heat transfer, inertia, air drag, and viscoelasticity, is to solve Equations 11.13a–c numerically as an initial value problem, adjusting Λ until the downstream boundary condition is satisfied (a "shooting method"). Convergence is rapid. It is important to keep in mind that all functions are complex. It can be shown from the linearity that there is no loss of generality in taking the real part of $\varpi(0)$ equal to 1 and the imaginary part equal to 0. There will be an infinite set of values $\{\Lambda_i\}$ for

Table 11.1. *Characteristic values for low-speed, isothermal spinning of a*
Newtonian liquid

$\ln D_R$	D_R	Λ_1		Λ_2	
		Real	Imaginary	Real	Imaginary
1.0	2.72	−3.810	7.671	−6.395	18.332
2.0	7.39	−2.025	10.845	−4.310	26.075
2.95	19.10	−0.120	13.814	−2.347	33.907
3.006	20.21	0.0	13.989	−2.219	34.380
3.15	23.34	+0.309	14.437	−1.884	35.596

which nontrivial solutions exist. The first two in magnitude are shown in Table 11.1.*
There is a sign change to a positive real part at a value of $D_R = 20.2$ (ln $D_R \sim 3$),
indicating that isothermal, low-speed spinning of a Newtonian fluid will be unstable
at draw ratios greater than that value. We cannot conclude from this analysis that
the instability will take the form of the sustained oscillations characteristic of draw
resonance, but that is a reasonable presumption and it is verified by nonlinear analyses, including direct numerical solution of the full transient Equations 11.6a–c. The
experimental transition to draw resonance for fluids that are Newtonian or nearly
Newtonian is close to a draw ratio of 20, but the precise value is uncertain because
of the ambiguity in defining the initial velocity and the likelihood of deviations from
the assumed kinematics near the takeup roll.

The analysis can be repeated with the energy equation included, as well as with
inertia, air drag, gravity, surface tension, and a purely viscous or viscoelastic non-
Newtonian constitutive equation. Cooling greatly increases the critical draw ratio,
as does inertia; these effects are probably the major reasons that commercial spinlines rarely experience draw resonance. Shear thinning decreases the critical draw
ratio substantially; the critical draw ratio for isothermal, low-speed spinning of an
inelastic power-law fluid with $n = 0.6$ is 8, while the critical value is less than 3 for
$n = 0.33$. Melt elasticity is stabilizing.

Figure 11.3 shows the critical draw ratio computed as above as a function of
$De = \lambda v_o/L$ for a single-mode Phan-Thien/Tanner liquid with $\varepsilon = 0$ and 0.015 and
various values of ξ. Line 1, with $\varepsilon = 0$ and $\xi = 0$, is a Maxwell fluid. The curve is
double valued for each De, indicating that there is a second stable region at very
high draw ratios, and no instability to infinitesimal disturbances can occur at any
draw ratio when De is greater than about 0.01. The double-valued curve is, to a
large extent, an artifact of the Maxwell fluid and parallels the "unattainable" region
defined by the line $D_R = 1 + De^{-1}$ (cf. Equation 10.6). The physical explanation
within the context of the constitutive model is that the spinline stress becomes so

---

* There is an important technical detail that will not concern us here. This system of equations is
not self-adjoint, so we do not have a proof that the complex eigenvalue with the smallest absolute
value will be the first to undergo a sign change in the real part. Thus, in principle we would have
to examine the entire set $\{\Lambda_i\}$ to determine the transition. Fortunately, the eigenvalue with the
smallest absolute value does seem to be the one that determines spinline stability. This is not a
general property of non–self-adjoint systems.

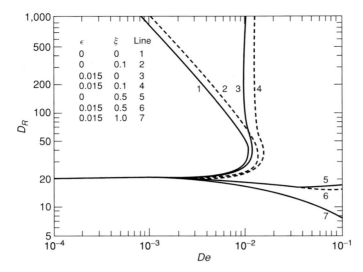

Figure 11.3. Critical draw ratio as a function of $De = \lambda v_o/L$ for a single-mode Phan-Thien/Tanner fluid with various values of ε and ξ. Reprinted from Chang and Denn, *Proc. 8th International Congress on Rheology*, Naples, Italy, Vol. 3, 1980, p. 9.

large at draw ratios approaching $1 + De^{-1}$ that disturbances cannot overcome the resistance to deformation and are thus unable to grow. There is only a small double-valued region for $\varepsilon = 0.015$ and $\xi = 0$, which is a fluid with a constant viscosity but bounded tensile stresses; these are the PTT parameters used in the polyester pilot plant spinline simulation in Figure 10.5. Line 7 of Figure 11.3, with the unusually large value $\xi = 1$, shows the destabilizing effect of shear thinning. The behavior is qualitatively the same for a two-mode PTT calculation.

The flat film process, in which an extruded sheet is drawn up on a roll, is a two-dimensional analog of melt spinning, and the dynamical equations and instabilities are essentially the same. The finite aspect ratio of the film die and the presence of an edge on the extruded sheet introduce stresses that cause necking in of the sheet. This is a steady-state phenomenon that does not seem to have major dynamical implications.

11.4 Sustained Oscillations and Draw Resonance

Linear stability theory can show definitively that a system is unstable, but it gives no information about the ultimate fate of the process as the disturbance grows. Furthermore, linear stability theory can show only conditional stability. There are two ways to attack the problem of finite disturbances. One is direct numerical simulation of the full set of nonlinear partial differential equations. This approach has become increasingly popular as computer power has grown, but a fundamental difficulty of distinguishing physical from numerical instability is always present. The other, employed less now than in the past, is to expand the nonlinear equations in

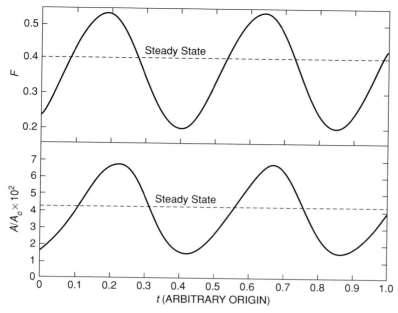

Figure 11.4. Sustained oscillations in force and area for low-speed spinning of a Newtonian liquid, $D_R = 23.34$. Reprinted from Fisher and Denn, *Chem. Eng. Sci.*, **30**, 1129 (1975).

an appropriate set of spatial functions with time-dependent coefficients and then to truncate in some predetermined manner to obtain equations for the coefficients; the expansion is usually done using the characteristic functions of the linear problem (the set $\{\psi_n, \phi_n, \varpi_n\}$ corresponding to the set of characteristic values $\{\Lambda_n\}$ in the spinline example, for instance).

Both approaches have been applied to low-speed spinning of a Newtonian fluid, with equivalent results. The conclusion is that the spinline is stable to finite disturbances below a draw ratio of 20.21, while at higher draw ratios the system is unstable and approaches a *limit cycle* characterized by sustained oscillations of the force and takeup area. Figure 11.4 shows the sustained oscillations computed by the expansion method with truncation at one term for $D_R = 23.34$, using Galerkin's method to obtain the ordinary differential equations for the real and imaginary parts of the area perturbation. This method is effective only for draw ratios close to the critical value because of the need to keep the number of terms in the expansion small, whereas direct numerical solution can produce wave forms like those in Figure 11.1.

11.5 Spinline Sensitivity

The classical approach to process sensitivity, which is the approach used in courses on process dynamics and control, is through *transfer function* analysis. The transfer function contains the response of the output to small input disturbances at all frequencies, and a knowledge of the transfer function is equivalent to knowledge of the dynamical response of the system to small but arbitrary disturbances. (The disturbances must be sufficiently small to permit the use of the linearized system equations.) Linear viscoelasticity, discussed in Section 9.2, is an example of transfer function analysis; $G'(\omega)$ and $G''(\omega)$ reflect the in-phase and out-of-phase responses to the sinusoidal forcing, and $|G^*(\omega)|$ is the amplitude ratio, or the magnitude of the ratio between input and output. Classical transfer function analysis typically uses the amplitude ratio and the phase angle of the response as the primary variables, rather than the separate values of the real and imaginary parts as is generally done in linear viscoelasticity. Transfer function analysis is nothing more than a method of converting the system description from the time domain to the Fourier transform (frequency) domain.

A linear system that is forced sinusoidally will have a sinusoidal response at the same frequency. One way to approach the analysis is then to assume an input of the form $\exp(i\omega t) = \cos(\omega t) + i \sin(\omega t)$, where $i^2 = -1$; the response will also have the form $\exp(i\omega t)$. The linearized equations for isothermal, low-speed Newtonian spinning, for example, Equations 11.10a–c, will then take the form of Equations 11.13a–c, with Λ replaced by $i\omega$; the functions ϕ, ψ, and ϖ are complex and are in fact the normalized Fourier transforms of A, v, and σ, respectively. The boundary conditions, however, are no longer zero, but reflect the forcing; if we wish to determine the sensitivity of the output area to disturbances in the velocity at $z = 0$, for example, we would set $\psi = 1 + 0i$ at $\zeta = 0$. (The input condition is the Fourier transform of an impulse, or a "delta" function, not a sinusoid, because the transfer function is the ratio of output to input in Fourier space. There is no loss of generality in setting the imaginary part to zero at $\zeta = 0$.)

Figure 11.5 shows the amplitude ratio and phase angle of the transfer function relating the takeup area to inlet velocity perturbations for isothermal, low-speed spinning of a Newtonian liquid. The conditions used for the calculation were $v_o = 13.2$ m/min and $L = 1$ m. The residence time, $t_R = \int dz/v(z)$, is $L(D_R - 1)/v_o D_R \ln D_R$ for isothermal Newtonian spinning and varies between 1.57 and 1.36 s for draw ratios between 15 and 25. Thus, $2\pi/t_R$ is about 4 rad/s. We focus first on $D_R = 15$, which is below the onset of draw resonance. The amplitude ratio approaches unity at low frequencies; this is simply steady-state conservation of mass, which requires a unit change in relative area for a unit change in relative velocity at a fixed throughput. The amplitude ratio decays at high frequencies, reflecting the fact that the system is unable to respond to disturbances that are much faster than the residence time. There is a resonant peak at about 3 rad/s, indicating that velocity disturbances in this range will be greatly amplified. What appears to be at work here is a reinforcement mechanism resulting from "reflection" of the velocity perturbation at the downstream boundary, where the takeup velocity is fixed; this is the reason that the phenomenon is seen at a frequency close to $2\pi/t_R$. The sequence of decaying resonant peaks is characteristic of hyperbolic (wave propagating) systems with reflecting boundaries; the transfer function of a countercurrent heat exchanger also has this appearance, for example. The amplitude ratio for $D_R = 25$, where this

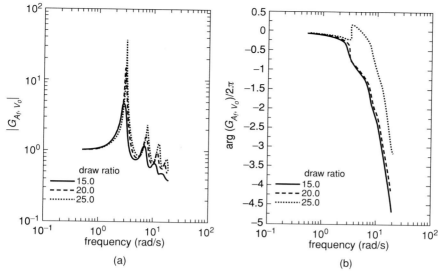

Figure 11.5. Transfer function between final area and inlet velocity for isothermal, low-speed spinning of a Newtonian liquid. (a) Amplitude ratio; (b) phase angle. Reprinted from Devereux, *Computer Simulation of the Melt Fiber Spinning Process*, Ph.D. dissertation, U. California, Berkeley, 1994.

system is known to exhibit draw resonance, is qualitatively the same, but we see very different behavior in the phase angle, which increases by 2π near the resonant frequency. This is an indication that the amplitude ratio has passed through infinity (all disturbances are amplified infinitely and the system runs away). Thus, stability information is contained in this analysis, but in a subtle way. What we really seek with this methodology is the information about sensitivity.

The issue of downstream boundary conditions becomes a bit more complex when heat transfer and solidification are included, since the solidification point will move dynamically when the system is perturbed. The transform methodology requires that the frequency response equations be solved on a fixed spatial domain. We can retain this structure as long as we retain the simplified condition that solidification occurs when the temperature reaches a fixed value; in that case the dynamics can be linearized about the solidification point, and the downstream boundary condition can be written in terms of a linear combination of the velocity and temperature perturbations at the steady-state solidification point. (This issue does not arise with the approach used by McHugh and co-workers, in which crystallization kinetics are included and the fixed domain is the entire spinline.)

Figure 11.6a shows a Newtonian fluid calculation of the amplitude ratios for the transfer functions of the final area relative to a variety of input disturbances for PET extruded at 295 °C from a 0.5-mm diameter spinneret at a velocity of 13.2 m/min and a draw ratio of 100. The cross-flow air had a velocity of 0.2 m/s and a temperature of 30 °C. Solidification was assumed to occur at 70 °C. Inertia and air drag were included in the steady-state and linearized equations, and perturbations were permitted in the temperature and velocity of the cross-flow air (T_a and v_a,

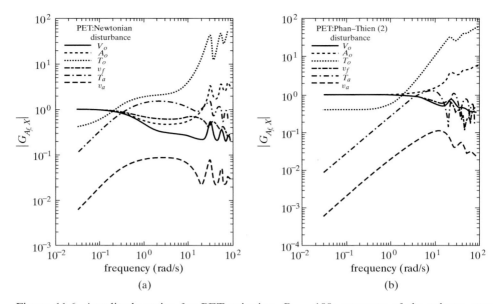

Figure 11.6. Amplitude ratios for PET spinning, $D_R = 100$; response of the takeup area to various input disturbances. (a) Newtonian fluid; (b) two-mode Phan-Thien/Tanner fluid. (Note the different vertical scales on the two figures.) Reprinted from Devereux, *Computer Simulation of the Melt Fiber Spinning Process*, Ph.D. dissertation, U. California, Berkeley, 1994.

respectively), as well as in the inlet and takeup parameters. Large amplitude ratios and resonant peaks are seen at frequencies corresponding to the residence time, but disturbances on this time scale ($\sim$0.2 s) are not possible for most of the variables. Only the cross-flow air velocity, which contains turbulent fluctuations, can possibly generate perturbations in a frequency range where this system can respond dynamically. Hence, we can conclude that the most likely source of diameter fluctuations will be fluctuations in the rate of heat transfer. The same calculation is shown in Figure 11.6b for a two-mode Phan-Thien/Tanner fluid with $\xi = 0.15$, $\varepsilon = 0.20$, $\lambda_{1,2} = 0.120 \times 10^{-3}$ and 0.673×10^{-2} s, respectively, and $G_{1,2} = 1.32 \times 10^{6}$ and 1.64×10^{4} Pa, respectively. As noted in Section 10.3, these parameters are a good fit to Gregory's rheology data and to George's spinline data. The overall response is similar to the Newtonian fluid, although the peaks are shifted a bit and some are modulated.

We now turn to an experimental study of spinline sensitivity to cross-flow air by Young and Denn (1989). The system is shown schematically in Figure 11.7. A melt filament was perturbed locally by cross-flow air variations, and perturbations in the drawn filament diameter were recorded. Experiments were carried out for PET and polypropylene. The spinneret diameter was 1 mm, and extrusion velocities ranged from 0.12 to 0.048 m/s. Draw ratios ranged from 52 to 180. The maximum takeup velocity was 450 m/min, so inertia and air drag were negligible. The experimental amplitude ratios are shown in Figure 11.8 for two PET runs, with $v_o = 0.12$ m/s and $D_R = 52$ and 68, respectively; the water bath was located 1.32 m from the spinneret for the data in Figure 11.8a and 0.70 m from the spinneret for the data

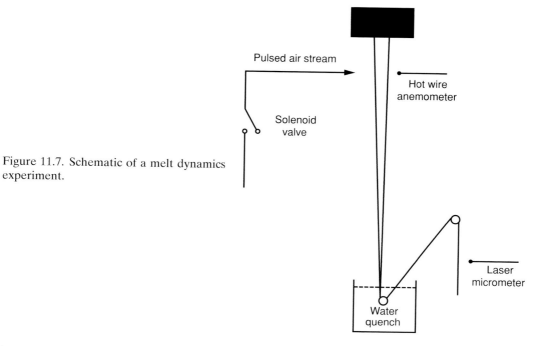

Figure 11.7. Schematic of a melt dynamics experiment.

in Figure 11.8b. Calculated amplitude ratios for Newtonian, Maxwell, and Phan-Thien/Tanner equations, using both one and two modes, are also shown. These runs represent the best and poorest agreement between the transfer function calculations and the PET data. There is essentially no difference between the PTT and Maxwell fluid calculations, nor between the use of one and two modes, although there is a difference between one and two modes in the steady-state calculations. The viscoelastic calculations do differ from the Newtonian. The location of the first resonant peak is captured, and the maximum gain is approximately correct, although the viscoelastic calculations overestimate the maximum sensitivity. The sharp minima in the computed amplitude ratios do not appear in the data, however. A large amount of "frequency cascading," a nonlinear phenomenon in which power is transferred to higher harmonics, was observed experimentally, as was some "smearing" of the output fundamental into nearby frequencies. Hence, the absence of the sharp minima and the overprediction of the magnitude might be caused in part by nonlinear effects.

Young and Denn also did a series of spinning experiments with a polypropylene melt. Polypropylene is highly shear thinning, and the polymer crystallizes on solidification. The melt data were fit with two- and four-mode PTT equations, with $\xi = 0.12$ and $\varepsilon = 0.035$. Spinning conditions were similar to the PET experiments, except that the PP was extruded at 200 °C. Two typical comparisons between the experimental and computed amplitude ratios are shown in Figure 11.9. The agreement here is completely unsatisfactory. The calculated frequency response underestimates the amplitude ratio by a considerable amount and fails badly in predicting the first resonant peak. In two of the five cases, the frequency response calculations

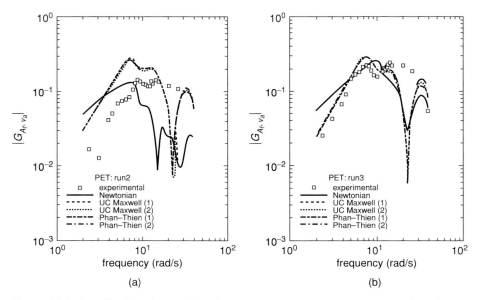

Figure 11.8. Amplitude ratio as a function of frequency for the response of the takeup area on a PET spinline to local perturbations in the cross-flow air velocity, Young's Runs 2 and 3. Reprinted with permission from Devereux and Denn, *Ind. Eng. Chem. Res.*, **33**, 2384 (1994). Copyright American Chemical Society.

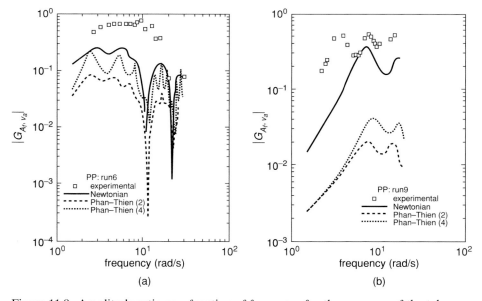

Figure 11.9. Amplitude ratio as a function of frequency for the response of the takeup area on a polypropylene spinline to local perturbations in the cross-flow air velocity, Young's Runs 6 and 9. Reprinted with permission from Devereux and Denn, *Ind. Eng. Chem. Res.*, **33**, 2384 (1994). Copyright American Chemical Society.

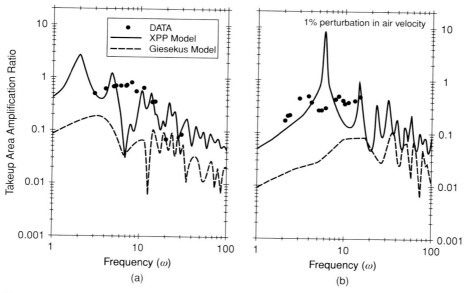

Figure 11.10. Amplitude ratio as a function of frequency for the response of the takeup area on a polypropylene spinline to local perturbations in the cross-flow air velocity. Young's Runs 6 (a) and 9 (b). Reprinted with permission from Kohler and McHugh, *Chem. Eng. Sci.*, **62,** 2690 (2007). Copyright Elsevier.

indicated that the spinning should be unstable and exhibit draw resonance, whereas draw resonance was not observed in the experiments.

Figure 11.10 shows a simulation of the same data by Kohler and McHugh, using a model in which stress-induced crystallization occurs and the solid phase carries a proportional fraction of the stress. The methodology for the sensitivity analysis is identical to that of Devereux and Denn except that the fixed spatial regime is the entire spinline and the downstream boundary condition is simply a zero perturbation in the takeup velocity. The melt stress was described by the Giesekus model, Equation 9.26, and by the XPP (extended pom-pom) model, which is a differential model based on the tube concept that has been used effectively in other studies to describe melt flow behavior. Kohler and McHugh's calculations indicated that there was little crystallization in Run 6, so direct comparison with the calculations in Figure 11.9a is appropriate (although they noted that only a small amount of crystallization is required to "lock in" the stresses). The response of the Giesekus model is essentially the same as the response of the Phan-Thien/Tanner model and is equally unsatisfactory. The XPP model is closer to the data in magnitude, although the location of the computed first resonant peak is also inconsistent with the data.

Kohler and McHugh's calculations for Young's Run 9 indicated that there was substantial crystallization. The Giesekus model for the melt is still inadequate, despite the inclusion of stress-induced crystallization in the spinline model, and the response does not differ greatly from the PTT simulations without crystallization.

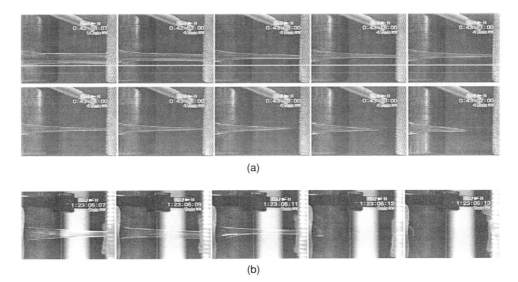

(a)

(b)

Figure 11.11. Failure of a polyisobutylene melt in extension. (a) Necking failure, stretch rate $= 0.058$ s^{-1}. (b) Cohesive failure, stretch rate $= 5.6$ s^{-1}. Courtesy of V. C. Barrosa and J. M. Maia.

The XPP model is again closer to the data in magnitude, but the first resonant peak is displaced and has an amplitude ratio that is a factor of 10 greater than the experimental amplification.

The conclusion based on these very limited dynamical studies is that existing models are adequate for understanding the spinline dynamics of amorphous polymers, such as PET at takeup speeds below 4,000 m/min, but they must be used with caution for semicrystalline polymers. It is not possible to draw a firm conclusion about the importance of the treatment of the solidification process in the frequency response analysis, although stress-induced crystallization and the effect of the crystalline phase on the mechanics are likely to be important and should be included since the theoretical framework is available.

11.6 Spinline Failure

Spinline failure is the primary dynamical problem in commercial spinning, and it is unlikely to be related to the instability and sensitivity mechanisms discussed in the preceding sections. The failure of polymer melt filaments in extensional flow occurs by three distinct mechanisms in different stretch rate regimes: surface tension-driven growth of disturbances that causes breakup into droplets, local necking that is analogous to ductile failure in solids, and cohesive failure that is analogous to brittle failure in solids. In addition, the presence of impurities, such as cross-linked gel particles, can cause failure. Figure 11.11 shows the failure of a polyisobutylene melt stretched at two different rates, illustrating necking at a low stretch rate and cohesive failure and substantial snapback at a high rate.

Surface tension–driven breakup into droplets is rarely important in melt spinning, where the large viscous and elastic forces overwhelm the surface tension forces. It is an important mechanism in the formation of the dispersed phase in polymer blends, and it is important in solution processing. The surface tension–driven breakup of a viscoelastic filament has been analyzed using both thin filament equations and a transient finite element analysis, but we will not pursue the topic here because it is not relevant to our present discussion.

The onset of ductile failure in solids is determined by the *Considère construction*, in which a maximum in the stress–strain curve causes an instability that manifests itself as a neck. This concept is unlikely to be applicable to the onset of necking in polymer melts. All constitutive equations, including the Maxwell model, Equation 9.16, predict a maximum in the stress–strain curve for stretching at a constant stretch rate, and this maximum normally occurs prior to the attainment of steady state. Hence, literal interpretation of the construction as a sufficient condition for failure would imply that uniform uniaxial extensional experiments could never be carried out past the force maximum, which often corresponds to a relatively low strain; such an interpretation is clearly contrary to substantial experimental experience in extensional rheometry, and several experimental studies focusing specifically on the Considère construction have shown that it does not predict the experimental onset of necking in melts.

Reiner and Freudenthal proposed a theory of cohesive failure for Maxwellian liquids in 1938; the theory requires a material "strength" with dimensions of a stress, but it reduces to the notion that there is a critical stress for failure that is characteristic of the material. The time to break for a given stretch rate is then calculated from Equation 9.22. In this approach, the stress and strain at failure are independent of the stretch rate, which is contrary to later careful experiments. There is no adequate theoretical treatment of cohesive failure within the context of the continuum models that we have been discussing.

Joshi and Denn have developed a molecular scaling argument for cohesive failure that is based on the tube concept. The model predicts a critical value of the recoverable strain, which is the amount of strain that is recovered by recoil following failure. The total strain at failure can then be calculated from a stress constitutive equation for a given recoverable strain. The basic concept is that an entangled polymer chain experiences a tension that is balanced by the friction with the surrounding chains with which it is entangled. The tension and the friction scale differently with chain extension, and there is a critical chain extension beyond which the two forces cannot balance. This point is interpreted as the onset of failure. All model parameters are derived from molecular structure or independent measurements, so the only adjustable parameter is a scaling coefficient that must be of order unity. Data in which both the strain at failure and the recoverable strain are available are very limited; the best data set is from the Vinogradov group in Russia in the 1970s. Vinogradov's failure data for a polyisoprene with a molecular weight of 575,000 and a polydispersity of 1.02 are shown in Figure 11.12, together with the calculated lines. (We_{rep} is the product of the stretch rate and the *reptation time*, which is the

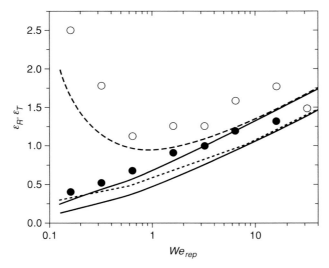

Figure 11.12. Failure data of Vinogradov for polyisoprene. The fine continuous and broken lines are the critical recoverable and total strain, respectively, from the scaling theory with a scaling coefficient of unity. The heavy continuous and broken lines are the critical recoverable and total strain from the scaling theory with a scaling coefficient of 2.5. Reprinted from Joshi and Denn, *J. Rheol.*, **48**, 591 (2004).

time scale associated with diffusion of the chain through the "tube.") The agreement is reasonably good if the scaling coefficient is set to 2.5, and it is likely that the scaling argument captures the essential physical phenomena. Scaling arguments are inherently crude, however, and a fundamental treatment of cohesive failure is still lacking.

11.7 Film Blowing

The blown film process is known to be difficult to operate, and a variety of instabilities have been observed on experimental and production film lines. We showed in the previous chapter (Figure 10.10) that even a simple viscoelastic model of film blowing can lead to multiple steady states that have very different bubble shapes for the same operating parameters. The dynamical response, both experimental and from blown film models, is even richer. The dynamics of solidification are undoubtedly an important factor in the transient response of the process, but the operating space exhibits a variety of response modes even with the conventional approach of fixing the location of solidification and requiring that the rate of change of the bubble radius vanish at that point.

Housiadas and co-workers have carried out a rigorous derivation of the blown film equations without crystallization for a class of constitutive equations that includes the Maxwell and PTT equations. They permitted nonaxisymmetric disturbances in the perturbation equations when studying stability, which had not been done in the small number of earlier studies. Results for a PTT fluid with $\varepsilon = 0.005$ and $\xi = 0$ are shown in Figure 11.13 for $Z/R_0 = 7$ and $\lambda V_0/R_0 = 0.1$.

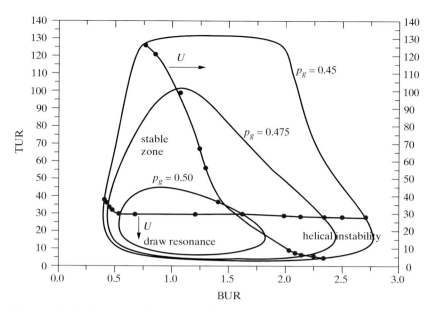

Figure 11.13. Contours of constant dimensionless pressure and regions of instability for a PTT fluid with $\varepsilon = 0.005$, $\xi = 0$, $Z/R_0 = 7$, and $\lambda V_0/R_0 = 0.1$. Reprinted with permission from Housiadas et al., *J. Non-Newtonian Fluid Mech.*, **141**, 193 (2007). Copyright Elsevier.

Here, $TUR = $ (thickness reduction)$/BUR$ and $p_g = 2B$ in terms of the nomenclature used in Section 10.4; TUR and BUR are the takeup ratio and blowup ratio, respectively. The region marked "U" is unstable according to the linear theory. Only results for $BUR > 1$ are of practical interest. We see that both draw resonance and helical instabilities can occur, with the latter comprising a large part of the parameter space for dimensionless pressure B less than about 0.25 ($p_g < 0.50$). This is consistent with what is seen experimentally, although the comparisons are qualitative.

Henrichsen and McHugh applied their model, which includes crystallization, to study both the sensitivity and the stability of the blown film process with axisymmetric disturbances. The sensitivity analysis showed no large resonant peaks in excess of the steady-state gain (the low-frequency asymptote of the amplitude ratio) like those for the spinline in Section 11.5. They did observe some effect of crystallization when the crystallization mechanics were "turned on" and "turned off" for otherwise identical conditions. The stability results show differences between parameters appropriate to low-density polyethylene and linear low-density polyethylene in a direction that is consistent with some experiments but not with others. Henrichsen and McHugh examined a larger range of BUR but a much smaller takeup ratio range than Housiadis and co-workers, and they did not include pressure contours, so it is not possible to make direct comparisons between the two studies to ascertain the importance of the crystallization model. What is needed is the incorporation of a crystallization model into a detailed study with nonaxisymmetric disturbances like that done by Housiadas and co-workers.

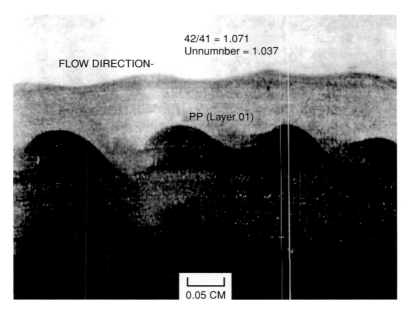

Figure 11.14. Photograph of a cross section of an extruded bicomponent film, with a polypropylene (Layer 1, above) and a high-density polyethylene (Layer 2, below). The flow direction is from left to right. The initial thickness ratio PP/HDPE was 1.071. Reprinted with permission from Wilson and Khomami, *J. Rheol.*, **37**, 315 (1993).

11.8 Interfacial Instabilities

Multilayer systems include bicomponent fibers and films with as many as several hundred coextruded layers. Steady interface motion that is induced in noncircular channels by the second normal stress difference was described in Section 10.6. The more common reason for interface distortion is the growth of interfacial instabilities as the film traverses the die. Figure 11.14 shows an extreme case of the development of interfacial waves in the coextrusion of a two-layer film consisting of a polypropylene (above) and a high-density polyethylene (below). The flow direction is from left to right. The interfacial disturbance was initiated by subjecting the flow to pressure fluctuations. The initial disturbance growth was exponential, as required by linear stability theory, but the crested waves that we see in the photograph are well beyond the linear regime.

The analysis of the growth of interfacial instabilities using a framework like that in Section 11.3 is relatively straightforward, but it is tedious because the linearized momentum and constitutive equations in both phases must be considered, together with a linearized equation for the interfacial dynamics. It is sometimes possible to obtain analytical solutions to interfacial instability problems when the characteristic wavelength can be taken to be long. Instability modes can be traced to differences in viscosity and normal stresses across the interface, and relative layer thickness is important. Linear stability analyses have been carried out for the multimode Giesekus fluid, as well as a single-mode variant of the Phan-Thien/Tanner model. The phenomenon, which includes "subcritical" instabilities in which nonlinear disturbances grow in a region where linear theory predicts stability to infinitesimal disturbances, is very complex, but the linear stability theory for the four-mode Giesekus fluid is in reasonable agreement with the data for interfacial disturbance growth.

11.9 Concluding Remarks

With the exceptions of extrusion die instabilities, which are addressed in the next chapter, instabilities in melt spinning and film blowing, and interfacial instabilities, the experimental literature on viscoelastic flow instabilities has emphasized dilute polymer solutions. The earliest studies were on the instability of flow between rotating concentric cylinders, known as the *Taylor-Couette* instability, where the steady in-plane rotational flow spontaneously breaks down into a series of superposed vortices. This is a well-studied inertial phenomenon for Newtonian fluids, and the inertial transition for dilute polymer solutions turns out to be surprisingly sensitive to the second normal stress difference. Giesekus observed a transition in this flow in 1962 at exceedingly small Reynolds numbers, but the observation appeared to be an anomaly; nearly three decades later, Larson, Muller, and Shaqfeh reported experiments and a linear stability analysis showing the existence of an "elastic" instability in this flow in the limit of zero Reynolds number. Other instabilities for creeping flow of viscoelastic liquids have also been observed, including torsional cone-and-plate flow, a "lid-driven" cavity, stagnation flow, and a cylinder in a channel, and it is quite likely that these phenomena have relevance to polymer melt processing, although direct connections have not been made.

The common feature of these instabilities at zero Reynolds numbers seems to be the coexistence of normal stresses and curved streamlines. McKinley has argued that both the experimental and theoretical onsets of instabilities for this class of flows correlate with a criterion of the form $\lambda U/(\mathcal{R}H)^{1/2} > M_{crit}$. λ is a mean relaxation time, U is the characteristic velocity, H is the length scale defining the shear rate, and $\mathcal{R}$ is the length scale defining the curvature of the streamlines. M_{crit} is a constant characteristic of the specific flow. U/H defines the shear rate and $\mathcal{R}/U$ the residence time, so the critical value can be interpreted as the geometric mean of the Deborah and Weissenberg numbers.

BIBLIOGRAPHICAL NOTES

The methodology for stability analyses is developed in detail in specialized texts. A broad introductory treatment can be found in

Denn, M. M., *Stability of Reaction and Transport Processes*, Prentice Hall, Englewood Cliffs, NJ, 1975.

Spinline stability is one of the examples in this book. The classic texts on fluid mechanical stability are

Chandrasekhar, S., *Hydrodynamic and Hydromagnetic Stability*, Oxford University Press, Oxford, 1961.

Lin, C. C., *The Theory of Hydrodynamic Stability*, Cambridge University Press, Cambridge, 1955; corrected printing, 1966.

There are two comprehensive reviews of stability in polymeric liquids,

Petrie, C. J. S., and M. M. Denn, *AIChE J.*, **22**, 209 (1976).
Larson, R. G., *Rheol. Acta*, **31**, 213 (1992).

The first focuses on polymer processing. Stability is addressed in many of the chapters in

Pearson, J. R. A., *Mechanics of Polymer Processing*, Elsevier Applied Science Publishers, London, 1985.

There is also a chapter on stability in

Tanner, R. I., *Engineering Rheology*, 2nd ed., Oxford University Press, Oxford, 2000,

and there are relevant chapters in

Hatzikiriakos, S. G., and K. Migler, Eds., *Polymer Processing Instabilities: Control and Understanding*, Marcel Dekker, New York, 2004.

The first analyses of draw resonance for an isothermal Newtonian fluid were done independently by Kase and co-workers and Matovich and Pearson in the mid-1960s. The early work on draw resonance and filament breakup is summarized in Petrie and Denn, cited above. Subsequent work is reviewed in

Denn, M. M., "Fibre Spinning," in J. R. A. Pearson and S. M. Richardson, Eds., *Computational Analysis of Polymer Processing*, Applied Science Publishers, London, 1983, pp. 179ff,

and the text by Pearson.

There has been a great deal of confusion in the spinning literature over the difference between sensitivity and stability, with the result that some utter nonsense has been published. The literature on dynamic simulation and process dynamics is surveyed in the chapter in *Computational Analysis of Polymer Processing* cited above; two elementary treatments from a traditional process dynamics point of view are

Kase, S., and M. M. Denn, *Proc. 1978 Joint Automatic Control Conf.*, Instrument Society of America, Pittsburgh, 1975, pp. II–71.
Denn, M. M., "Modeling for Process Control," in C. T. Leondes, Ed., *Control and Dynamic Systems,* Vol. 15, Academic Press, New York, 1979, pp. 147ff.

The development of the linearized dynamical boundary condition at solidification is in

Chang, J. C., M. M. Denn, and S. Kase, *Ind. Eng. Chem. Fundam.*, **21**, 13 (1982).

Transfer functions for a large number of inputs to a Newtonian spinline are in

Kase, S., and M. Araki, *J. Appl. Polym. Sci.*, **27**, 4439 (1982).

The frequency response experiments are from

Young, D. G., and M. M. Denn, *Chem. Eng. Sci.*, **44**, 1807 (1989).

The frequency response calculations are mostly from

Devereux, B. D., and M. M. Denn, *Ind. Eng. Chem. Res.*, **33**, 2384 (1994),

where references to other work on spinline dynamics may be found. Much of the detail of the frequency response analysis is in

Devereux, B. D., *Computer Simulation of the Melt Fiber Spinning Process*, Ph.D. dissertation, University of California, Berkeley, 1994.

The frequency response calculations in Figure 11.10, which include the crystallization model, are from

Kohler, W. H., and A. J. McHugh, *Chem. Eng. Sci.*, **62**, 2690 (2007).

Spinline failure is addressed in the 1976 review by Petrie and Denn cited above. There is a good review of failure in extension that focuses on Russian work from Vinogradov's laboratory in

Malkin, A. Ya., and C. J. S. Petrie, *J. Rheol.*, **41**, 1 (1997),

and the topic was also addressed in another review in the same year:

Ghijsels, A., C. H. C. Massardier, and R. M. Bradley, *Int. Polym. Proc.* , **12**, 147 (1997).

A more recent review is

Joshi, Y. M., and M. M. Denn, "Failure and Recovery of Entangled Polymer Melts in Elongational Flow," in D. M. Binding and K. Walters, Eds., *Rheology Review 2004*," British Soc. Rheology, 2004, p. 1.

The surface tension-driven breakup of a viscoelastic jet is analyzed in

Bousfield, D. W., R. Keunings, G. Marrucci, and M. M. Denn, *J. Non-Newtonian Fluid Mech.*, **21**, 79 (1986).

Application of the Considère criterion to the necking of polymer melts seems first to have been done by

Cogswell, F. N., and D. R. Moore, *Polym. Eng. Sci.*, **14**, 573 (1974).

A discussion of the available literature and the limitations of the approach for polymer melts is contained in the review by Joshi and Denn cited above.

The scaling theory of cohesive rupture is from

Joshi, Y., and M. M. Denn, *J. Rheol.*, **47**, 291 (2003).

The two cited studies of stability and sensitivity of film blowing are

Henrichsen, L. K., and A. J. McHugh, *Int. Polym. Proc.*, **XXII**, 190 (2007),
Housiadas, K. D., G. Klidis, and J. Tsamopoulos, *J. Non-Newtonian Fluid Mech.*, **141**, 193 (2007)

where references to a few earlier studies and the experimental literature may be found. There is consistency between most of the results in Housiadas et al. and the earlier study by Cain and Denn,

Cain, J. J., and M. M. Denn, *Polym. Eng. Sci.*, **28**, 1527 (1988),

but a transcription error in the latter that propagated through the stability equations makes the results unreliable.

The photograph of the interfacial instability is from

Wilson, G. M., and B. Khomami, *J. Rheol.*, **37**, 315 (1993).

For analyses and reviews of the earlier experimental and theoretical literature, see

Ganpule, H. K., and B. Khomami, *J. Non-Newtonian Fluid Mech.*, **81**, 27 (1999).
Renardy, Y. Y., and M. Renardy, *J. Non-Newtonian Fluid Mech.*, **81**, 215 (1999).
Khomami, B, and K. C. Su, *J. Non-Newtonian Fluid Mech.*, **91**, 59 (2000).

The article by Ganpule and Khomami contains physical arguments about the mechanisms for interfacial disturbance growth, as does

Huang, C.-T., and B. Khomami, *Rheol. Acta*, **40**, 467 (2001).

This last article is concerned with instabilities in multilayer films with a free upper surface.

Instabilities in viscoelastic liquids at low Reynolds numbers are reviewed in

Shaqfeh, E. S. G., "Fully Elastic Instabilities in Viscometric Flows," in J. L. Lumley, M. Van Dyke, and H. L. Reed, Eds., *Ann. Rev. Fluid Mech.*, Vol. **28**, Annual Reviews, Inc., Palo Alto, CA, 1996, pp. 129ff.

For subsequent studies of the Taylor-Couette instability, see

Baumert, B. M., and S. J. Muller, *J. Non-Newtonian Fluid Mech.*, **83**, 33 (1999).
Al-Mubaiyedh, U. A., R. Sureshkumar, and B. Khomami, *J. Rheol.*, **44**, 1121 (2000).
White, J. M., and S. J. Muller, *Phys. Rev. Lett.*, **84**, 5130 (2000).

McKinley's criterion for flow with curved streamlines and generalizations for more complex rheology are in

McKinley, G. H., P. Pakdel, and A. Öztekin, *J. Non-Newtonian Fluid Mech.*, **67**, 19 (1996).
McKinley, G. H., "Extensional Rheology and Flow Instabilities in Elastic Polymer Solutions," in M. J. Adams, R. A. Mashelkar, J. R. A. Pearson, and A. R. Rennie, Eds., *Dynamics of Complex Fluids*, Imperial College Press and the Royal Society, London, 1998, p. 6.

Subsequent work on lid-driven flows is in

Padkel, P., and G. H. McKinley, *Phys. Fluids*, **10**, 1058 (1998).
Grillet, A. M., E. S. G. Shaqfeh, and B. Khomami, *J. Non-Newtonian Fluid Mech.*, **94**, 15 (2000).

Stagnation flow is addressed in

Öztekin, A., B. Alakus, and G. H. McKinley, *J. Non-Newtonian Fluid Mech.*, **72**, 1 (1997).

Experiments on torsional flow and an overview of the onset and properties of "elastic turbulence," as these low Reynolds number instabilities are sometimes known, is in

Burghelea, T., E. Segre, and V. Steinberg, *Phys. Fluids*, **19**, 053104 (2007),

where references to an interesting body of work on dilute polymer solutions by Steinberg and co-workers may be found.

12 Wall Slip and Extrusion Instabilities

12.1 Introduction

The no-slip boundary condition is introduced in every text on fluid mechanics as a fundamental principle of the discipline, and students are expected to accept the condition as "obvious." No-slip is not obvious, however, nor was it obvious to the founders of the discipline of fluid mechanics in the nineteenth century. Navier, in 1823, believed that there should be a relative ("slip") velocity between the fluid and the solid surface that would be proportional to the shear stress at the wall; in modern terminology we would write Navier's hypothesis as (cf. Equation 2.30)

$$\eta v_s = b\tau_w,$$ (12.1)

where v_s is the slip velocity and τ_w is the shear stress. η is the viscosity, and b is a parameter with dimension of length that would presumably be a function of the fluid–solid pair. The no-slip condition for fully developed laminar tube flow of a Newtonian fluid predicts a fourth-power dependence of flow rate on tube radius, which is an extremely sensitive function. Careful experiments with water in round capillaries in the first part of the nineteenth century by Hagen and Poiseuille verified the fourth-power dependence, thus establishing the no-slip condition experimentally. Theoreticians, notably including Stokes, cautiously accepted the no-slip condition, and it became a part of the theoretical canon. Even into the mid–twentieth century, however, one finds occasional questions in the literature about the general validity of the boundary condition.

The development of the polymer processing industry following World War II renewed interest in the relevance of the no-slip boundary condition; this probably occurred because of the realization that polymer melts show rubberlike behavior on short time scales and rubber does not adhere to the wall during compounding. Benbow and Lamb, who were studying extrusion instabilities (which we address later in this chapter), reported slippage along the die wall in 1963 during the extrusion of a polyethylene; this was done by observing the motion of a colored dye, and they noted that the onset of unstable flow is affected by the materials of construction of the extrusion die. The materials of construction should have no effect on the flow if the no-slip condition is satisfied. Other experiments at about the same time also reported slip in melt flow by tracking tracer particles near the wall, but these experiments were discredited by calculations showing that velocity gradient across the finite particles could have caused the motion even with the no-slip condition, and the subject received little further attention for two decades. Experiments by Ramamurthy in 1986 showing that linear low-density polyethylene (LLDPE) melts exhibit very different flow behavior in geometrically identical extrusion dies fabricated from steel and α-brass renewed interest in possible slip flow in melts. The subject has since become of considerable interest to the microfluidics community as well, where the focus is typically on low molar-mass liquids, since the large surface-to-volume ratio in a microchannel greatly enhances the importance of any wall-region effects.

12.2 Experiments

It is now generally accepted that polymer melts exhibit apparent wall slip in flow at high stress levels, with an impact on the design and control of extrusion and molding flows. (We use the term *apparent wall slip* here to emphasize that most experiments are macroscopic and cannot distinguish between anomolous wall-region behavior that looks like slip and a true absence of adhesion at the molecular level. Whenever we use the terminology *wall slip* in this chapter, we mean apparent wall slip.) One common way to determine wall slip is to perform flow experiments in a series of capillaries of different radii, and then to plot the apparent shear rate $\dot{\gamma}_a = 4Q/\pi R^3$ versus $1/R$ at constant wall stress, where Q is the volumetric flow rate and R is the radius. It is straightforward to demonstrate that the slope of the line is four times the slip velocity. Results of Ramamurthy on an LLDPE in a steel die are shown in Figure 12.1; apparent slip velocities of order $1 - 10$ mm/s were observed. The flow curves by Ghanta and co-workers, shown in Figure 12.2, were obtained for an LLDPE in geometrically identical α-brass and stainless steel dies. The throughput is considerably higher in the brass die relative to the steel die at the same wall stress, indicating apparent slip. (The bars on the steel die data do not represent experimental uncertainty, which is smaller than the size of the data points. They reflect the range of certain pressure fluctuations that are discussed later in this chapter.) Laser-Doppler velocimetry measurements of a linear high-density polyethylene by Munstedt and co-workers are shown in Figure 12.3. These data clearly extrapolate to a nonzero velocity at the wall.

Wise and co-workers used an infrared (IR) evanescent wave spectroscopy technique to follow the disappearance of a tracer of deuterated 1,4-polybutadiene (PBDE) displaced in a channel by unlabeled PBDE of the same molecular weight. Linear PBDE is a convenient polymer for experimental research because it flows at room temperature and has a viscosity that is relatively insensitive to shear rate. One face of the channel was fabricated from a ZnSe IR crystal, and the deuterated polymer was placed on the surface. The deuterated polymer was then allowed to equilibrate with hydrogenated polymer of the same molecular weight, and flow was initiated. The concentration of deuterated polymer in the wall region after the

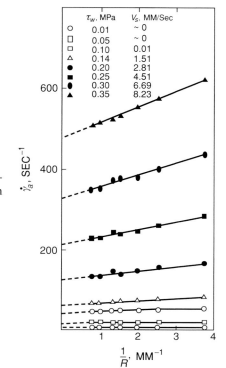

Figure 12.1. Apparent shear rate as a function of $1/R$ at constant wall stress for an LLDPE. Reprinted with permission from Ramamurthy, *J. Rheol.*, **30**, 337 (1986).

initiation of flow of the undeuterated polymer was monitored by the intensity of the IR signal, which was operated in "total reflection" mode to probe only the region near the wall. The signal intensity is proportional to concentration. It can be shown that normalized concentration is a unique function of the product $\dot{\gamma}_a^{2/3}t$ for no-slip flow with transverse diffusion of the species at the wall. Data for normalized

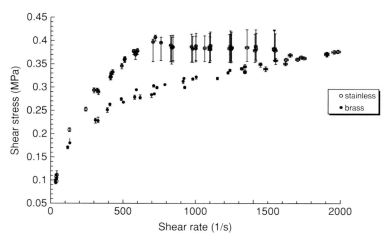

Figure 12.2. Shear stress as a function of shear rate for an LLDPE in geometrically identical steel and α-brass dies. Reprinted from Ghanta et al., *J. Rheol.*, **43**, 435 (1999).

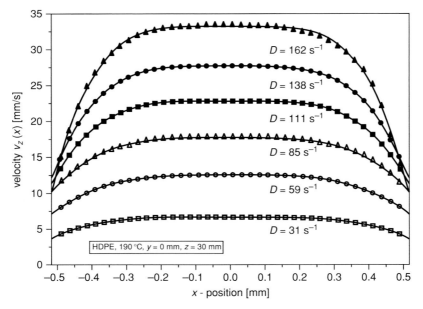

Figure 12.3. Laser-Doppler velocity profiles for an HDPE. Reprinted with permission from Munstedt et al., *J. Rheol.*, **44**, 413 (2000). *D* denotes the shear rate.

absorbance plotted as a function of $\dot{\gamma}_a^{2/3}t$ are shown in Figure 12.4; the data do superimpose at the lower stresses, consistent with the no-slip condition, but the increased decay rates at higher stresses indicate that the tagged material is leaving more quickly than would occur by no-slip flow and transverse diffusion; hence, there must be slip. The IR signal averages over a distance of several microns from the wall, so spatially resolved concentration profiles cannot be obtained, therefore precluding the possibility of determining the actual mechanics of slip. It is possible to conclude from the curvature of these data, however, that true slip (i.e., adhesive failure) at the die wall did not occur at these stresses.

The only available data on polymer melt flow near a wall with resolution approaching the size of an extended polymer chain are by Leger and co-workers, who used optical evanescent wave spectroscopy with photobleaching of fluorescent tracer molecules to carry out direct observations of the velocity within 100 nm of a surface. The results of their measurements in plane Couette flow of a linear polydimethylsiloxane (PDMS) with a molecular weight of about 10^6 past a silica surface containing terminally grafted PDMS chains with a molecular weight of about 10^5 are discussed below. Their studies suggest an apparent slip plane that is located away from the wall, with slip velocities for this system that are much smaller than those deduced for linear polyethylene by Ramamurthy and others using macroscopic methods.

The general conclusion that can be reached from these experiments and others not mentioned here is that apparent slip is observed for some highly entangled linear polymers, while it has not been observed for branched polymers or for linear polymers with an insufficient number of entanglements per chain. None of the

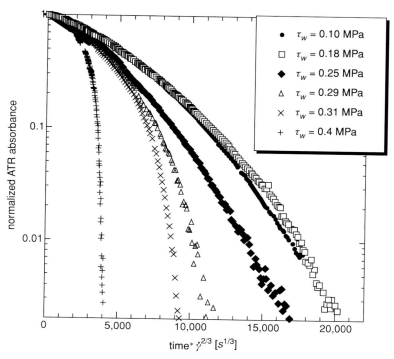

Figure 12.4. Normalized infrared intensity over a ZnSe surface, polybutadiene with $Mw = 180,000$. Reprinted from Wise et al., *J. Rheol.*, **44**, 549 (2000).

experimental methods has sufficient resolution to determine the mechanism of the apparent slip, although at least at modest stress levels it appears to involve a physical process that is operative away from the die wall.

12.3 Theories of Slip

There are three broad pictures of wall slip in polymer melts and concentrated solutions: (i) Slip is the result of an adhesive failure of the polymer chains at the solid surface, resulting in an interface that is polymer-free; (ii) slip is a cohesive failure resulting from disentanglement of chains in the bulk from chains adsorbed at the wall, resulting in an apparent "failure plane" within the polymer; and (iii) there is a low-viscosity lubricated layer at the wall, perhaps the result of a stress-induced transition to an ordered low-viscosity liquid crystalline mesophase or of stress-induced diffusion of smaller chains in the molecular weight distribution to the wall region. (Scenarios (ii) and (iii) are not necessarily distinct, since a consequence of disentanglement would be a layer of relatively unentangled polymer chains at the wall, which would have a lower viscosity than the entangled chains in the bulk.) The disentanglement picture is the most likely, except perhaps at the highest stresses, and that is the only one that we address here in any detail.

The notion that slip in a polymer melt is a consequence of disentanglement of chains adsorbed to the wall from those in the bulk seems to have been proposed first

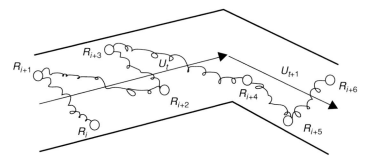

Figure 12.5. Schematic of the stochastic tube model.

by Bergem in 1976. Brochard and de Gennes developed a theoretical foundation in 1992 using scaling arguments. A number of subsequent groups have quantified the concept using the "tube" picture of entangled polymer chains mentioned in Section 9.5.4, and we briefly describe one such approach below. The tube theories are based on the notion that there are two types of polymer chains, those that are at the wall and those in the bulk. The chains at the wall are assumed to have a known density and to be tethered to the wall at one end. The tethering assumption is necessary for all formalisms, and it is consistent with the set of well-defined experiments on PDMS by Leger and co-workers that have been used for comparison with theory. It is only a crude approximation to the real extrusion situation, however, since polymers will adsorb to the wall all along the chain and not only at the end. Hence, chain "loops" are ignored, and the actual lengths of the "tails" that extend away from the wall are unknown. In addition, the density of adsorbed chains is unknown but is likely to depend on the surface chemistry; indeed, such a dependence on the surface chemistry is the only consistent way to explain the different behavior of LLDPE in brass and steel capillaries. (The fact that the chains are adsorbed and not permanently attached to the wall, as assumed in the theories, is probably of lesser importance.)

12.4 Tube Theory of Wall Slip

The starting point for the tube theory of wall slip employed by Xu and co-workers is shown in Figure 12.5. The transverse motion of a polymer chain in a melt or concentrated solution is restricted by the surrounding chains, so the primary diffusive motion must be along the backbone. (This motion is known as *reptation*, indicating a snakelike motion.) The surrounding chains are taken in a mean-field sense to comprise a *tube*, with a diameter that is on the order of the distance between entanglements. (*Entanglement* is an abstract but useful concept in polymer physics; no one knows precisely how polymer chains in a dense system hinder the motion of neighboring chains. The mean distance between entanglements is a known property of the linear molecule, however. The *entanglement molecular weight* is the molecular weight at which the dependence of the viscosity on molecular weight changes from linear to a power of about 3.4. The equilibrium number of entanglements per chain is simply the ratio of the molecular weight to the entanglement molecular weight.)

The hypothetical tube is a dynamical element that convects with the surrounding fluid. The changes of direction of the tube shown in the figure correspond to entanglement points in the model. The chain, which is represented by a series of friction points ("beads") connected by finitely extensible springs, moves within the constraints of the tube. The beads experience Brownian and frictional forces, as well as the tensile stress imparted by the stretched springs. There is also an osmotic-like force that is a consequence of the restriction of the chain to the tube; this force manifests itself with a tensile component that prevents chain collapse within the tube and a transverse component with a potential that is quadratic in deviations from the tube centerline. Application of Newton's second law to each bead results in a set of stochastic differential equations. (The equations are stochastic because the Brownian force is random, with a zero mean and a prescribed variance.) If an end of the chain retracts into the tube past an entanglement point, the entanglement is lost and the tube segment is removed. A new entanglement is formed and a new tube segment is created if an end of the chain moves out of the tube a distance equal to the equilibrium length between entanglements. The calculation is carried out for an ensemble of chains. Clearly, each entanglement must involve two chains (we ignore the possibility that a chain might entangle with itself), so when an entanglement is lost because of chain retraction the complementary entanglement on the other chain must also be lost. Similarly, a complementary entanglement must be created on another chain when a new entanglement is created because the chain has emerged from the tube. There are rules for these and other details, but they are not important in the present context. Stresses are computed from an ensemble average of an orientation tensor constructed from segment directions in the manner usually employed for statistical theories of polymer rheology.

The theory has only a single adjustable parameter, which corresponds to the *Rouse time* (the characteristic relaxation time for an unconfined chain) of the polymer, and it does a quite reasonable job of predicting the linear viscoelastic response and the transient and steady-state shear and normal stresses in simple shear. It is not as good as more complex tube-based models like the pom-pom model, and it cannot be used for nonviscometric flows because of the absence of a continuum representation, but it contains structural details and is very useful for providing insight into the mechanics of slip.

The application of the stochastic tube theory to slip requires that there be three types of chains: bulk chains that do not interact with the chains at the wall, wall chains that do not interact with the chains in the bulk, and free chains that interact with both the bulk and wall chains. The model is applied to each type of chain, and there are again details of implementation that need not concern us here. The wall stress is determined from the bulk behavior, and because the stress must be continuous, the shear rate experienced by the wall chains that is required to match the stress in the bulk is computed. The density and length of the wall-tethered chains must be known. The only experiments with a known surface density of wall-tethered

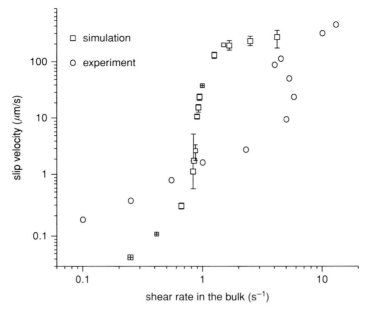

Figure 12.6. PDMS slip velocity data of Leger and co-workers and predictions of the stochastic tube model. Reprinted from Xu et al., *J. Rheol.*, **51**, 451 (2007).

chains with known molecular weight are those by Leger and co-workers on linear PDMS, which were mentioned above. The slip velocity data are shown in Figure 12.6, together with the predictions of the tube theory, where the only adjustable parameter was the Rouse time of the bulk melt. The model captures the features and magnitude of the slip data, except for the small amount of hysteresis in the experimental data. The difference between theory and experiment in the shear rate at which the large increase in slip velocity occurs is within the range of uncertainty of the Rouse time, which enters as a time-scaling parameter, but it is more likely that the difference is a consequence of the simplicity of the model. The calculations indicate that the density of entanglements in the wall region becomes significantly less than in the bulk, resulting in a wall layer of relatively unentangled chains with a viscosity lower than that in the bulk. Thus, the results of the theory are consistent with Scenario (ii) above.

12.5 Slip In Flow Simulation

Incorporation of an empirical slip boundary condition in simulation is straightforward, and we noted in the flow examples in Chapter 8 that slip was introduced as a means of smoothing discontinuities in boundary conditions. Most slip data for polymer melts have been obtained in shear flow over a narrow range of shear rates, and the data can usually be fit with a power relation of the form $v_s = \text{constant} \times \tau_w^m$, where values of the exponent m have been reported in various studies to be in the range from 2 to 6. The Navier boundary condition corresponds to $m = 1$.

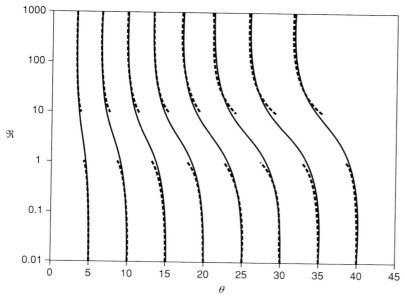

Figure 12.7. Computed streamlines for plane converging flow of a Newtonian fluid with a Navier slip boundary condition. Reprinted from Joshi and Denn, *J. Non-Newtonian Fluid Mech.*, **114**, 185 (2003).

The consequence of a slip boundary condition will be felt the most in a flow with a changing cross section, since in that case there may be large changes in the local wall stress at a fixed throughput. This point is nicely illustrated by flow between infinite converging planes, which was used as an illustrative example in Chapter 2. We showed there that radial streamline flow is possible in the creeping flow limit for a Newtonian liquid only under conditions of no slip or perfect slip. The same result easily follows for inelastic shear-thinning fluids. (Radial streamline flow is not possible in this geometry in general for viscoelastic liquids and a no-slip boundary condition, although the deviation from radial flow takes place primarily near the exit, where the radial assumption probably breaks down in any event.) We can visualize the nature of the flow that we expect when slip is possible. Far from the exit, where the cross section is large, the flow is slow and the wall stress is small; here the no-slip condition will approximately apply and the flow will be radial, with a maximum in the radial velocity at the center plane. Close to the exit the cross section is small, the flow is rapid, and the wall stress is large; here the perfect slip condition will approximately apply and the flow will be radial, with a uniform radial velocity. Hence, there must be a transition region with a transverse velocity in order to effect the change between the two limiting velocity profiles.

Computed streamlines in plane converging flow with a half-angle of $\pi/2$ for a Newtonian fluid with the Navier boundary condition are shown in Figure 12.7. The streamlines are shown on an orthogonal grid, where radial flow appears as a vertical line, to facilitate interpretation. The dimensionless radial position, denoted $\Re$, equals r/b, where b is the slip length defined in Equation 12.1. The solid lines are the

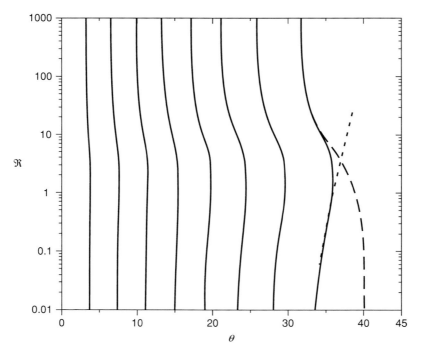

Figure 12.8.   Streamlines for a Carreau-Yasuda fluid with $n = 0.3$ and $w = 4.0$. The dashed line is the streamline for a Newtonian fluid ($w = 0$), while the dotted line is the streamline for the corresponding power-law fluid. Reprinted from Joshi and Denn, *J. Non-Newtonian Fluid Mech.*, **114**, 185 (2003).

results of a finite element calculation, while the dashed lines are asymptotic solutions for $\mathfrak{R} \gg 1$ (approaching no slip) and $\mathfrak{R} \ll 1$ (approaching perfect slip). The transition flow occurs for $\mathfrak{R} = O(1)$, where the distance from the point of convergence is comparable to the slip length b.

A similar asymptotic solution can be obtained for power-law fluids. One counterintuitive phenomenon with power-law behavior occurs when $n = \frac{1}{2}$, in that the flow is radial everywhere with partial slip; the downstream asymptotic behavior changes from nearly perfect slip to no slip as the power-law index n decreases below $n = \frac{1}{2}$.

A second dimensionless group, $w = q\beta/2(\alpha b)^2$, arises for the Carreau-Yasuda (C-Y) shear-thinning fluid, Equation 2.40b:

$$\eta = \eta_o \left(1 + \beta(\tfrac{1}{2}II_D)^{\frac{a}{2}}\right)^{\frac{n-1}{a}}. \tag{12.2}$$

The C-Y fluid approximates a power-law fluid with power-law index n at high shear rates and approaches a Newtonian fluid with a viscosity equal to η_o for low shear rates. Streamlines from a finite element simulation for a C-Y fluid are shown in Figure 12.8 for a power-law index $n = 0.3$ and $w = 0.4$. $\mathfrak{R}$ is defined as appropriate for a Newtonian fluid with the zero-shear viscosity of the C-Y fluid; in that case it can be shown that we expect deviations from Newtonian behavior when $\mathfrak{R}^2 \sim w$. The dashed line is the streamline for a Newtonian fluid ($w = 0$), while the dotted line

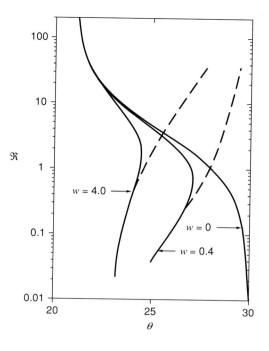

Figure 12.9. A streamline for a Carreau-Yasuda fluid with $n = 0.3$ for various values of w; the Newtonian fluid is $w = 0$. The dashed lines are the corresponding power-law streamlines. Reprinted from Joshi and Denn, *J. Non-Newtonian Fluid Mech.*, **114**, 185 (2003).

is the asymptotic solution for the power-law fluid with $n = 0.3$. The streamlines for the Newtonian and C-Y fluids are identical for $\mathfrak{R} \gg w^{1/2}$, while the C-Y streamlines merge into the $n = 0.3$ power-law streamlines for $\mathfrak{R} \ll w^{1/2}$. The azimuthal flow is always away from the center plane for $\mathfrak{R} < w^{1/2}$; it is toward the center plane for $\mathfrak{R} < w^{1/2}$, as shown here, when $n < 0.5$.

Figure 12.9 shows a single streamline for a C-Y fluid with $n = 0.3$ for various values of w. The two broken lines denoting the power-law fluid do not overlap because of the use of length scaling based on the zero-shear viscosity. The maximum deviation from the upstream no-slip Newtonian flow and downstream no-slip C-Y flow occurs for $\mathfrak{R} \approx w^{1/2}$, where the azimuthal velocity changes direction. The substantial streamline curvature exhibited by the C-Y fluid in a contraction forces one to think about the elastic instabilities generated by curved streamlines mentioned in the preceding chapter, and indeed this phenomenon may be related to extrusion instabilities that are believed to originate in the die entry region. We briefly discuss these instabilities later in this chapter.

12.6 Sharkskin

Some highly entangled linear polymers, including high-density polyethylene (HDPE), linear low-density polyethylene (LLDPE), 1,4-polybutadiene (PBDE), and polydimethylsiloxane (PDMS), exhibit an extrusion instability known as *sharkskin*, in which a small-amplitude, high-frequency disturbance appears on the extrudate surface at a wall shear stress that is typically on the order of the linear viscoelastic plateau modulus (0.1 – 0.2 MPa for most polymers). Figure 12.10 shows

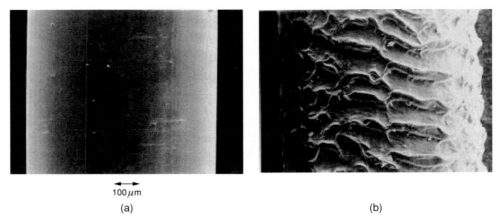

100 μm

(a)   (b)

Figure 12.10. Extrudates of an LLDPE at 145 °C at (a) 0.12 MPa and (b) 0.29 MPa, illustrating the onset of sharkskin. Reprinted from Pudjijanto and Denn, *J. Rheol.*, **38**, 1735 (1994).

extrudates of an LLDPE at 145 °C at (a) 0.12 MPa and (b) 0.29 MPa, where the latter illustrates the onset of sharkskin.

There was a widespread belief in the latter part of the twentieth century that the onset of sharkskin was in some way associated with the onset of wall slip, and the literature on the two subjects became inexorably intertwined. This association was based in part on the fact that observable wall slip occurs for these polymers at stresses of the same magnitude as those at which visible sharkskin occurs. It is now generally accepted that sharkskin is caused by a "tearing" (tensile) failure at the point where the melt exits the die; the periodicity is a result of dynamic crack penetration and subsequent healing, after which the process repeats. There is no quantitative theory for tensile failure, as noted in the previous chapter, but visual data appear to be definitive and there are (reasonably converged) finite element calculations showing that the tensile stress in the region of the point of departure from the die at the experimental onset of sharkskin is about the same as the stress that causes failure in extension for the same polymer. The onset of sharkskin can be delayed by the use of "flow modifiers," which are typically fluoropolymers that enhance slip, minimizing the amount of flow reorganization required in the neighborhood of the die exit, hence lowering the amount of stretching and the concomitant tensile stress for a given throughput. The slip flow caused by the use of a brass die has the same effect. Figure 12.11 shows extrudates corresponding to the data in Figure 12.2 at a wall stress of 0.295 MPa. The upper extrudate, which is smooth, is from the brass die at an apparent shear rate $8V/D$ of 589 s^{-1}, while the lower sharkskinned extrudate is from the steel die at $8V/D = 298$ s^{-1}.

12.7 Slip-Stick Flow

Linear polymers that exhibit sharkskin typically also exhibit a discontinuity in the shear flow curve known as *slip-stick*. The phenomenon is illustrated in Figure 12.12 for an LLDPE in a piston-driven capillary rheometer. At a critical wall stress there

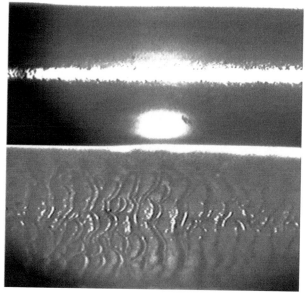

Figure 12.11. Extrudates of LLDPE at $\tau_w = 0.295$ MPa from (upper) an α-brass die at rate $8V/D = 589$ s^{-1} and (lower) a stainless steel die at $8V/D = 298$ s^{-1}. Reprinted from Ghanta et al., *J. Rheol.*, **43**, 435 (1999).

is a jump in the throughput, with a small amount of hysteresis on the return. The behavior is reminiscent of an ignition-extinction phenomenon in combustion, in which the system jumps discontinuously between states.

The data points in the region of discontinuity represent mean flow rates calculated from the velocity of the driving piston. In reality, the output oscillates between the upper and lower branches of the flow curve. The bars on the stainless steel data in Figure 12.2 denote the range of pressure fluctuations during these oscillations. (The pressure fluctuations are small relative to the base pressure, so they do not affect the flow curve in any substantive way.) There is massive slip on the

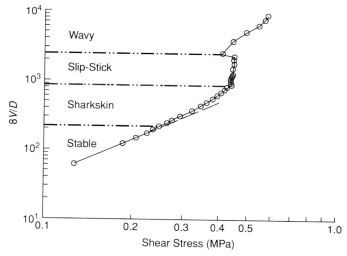

Figure 12.12. Flow curve of an LLDPE. Reprinted from Kalika and Denn, *J. Rheol.*, **31**, 815 (1987).

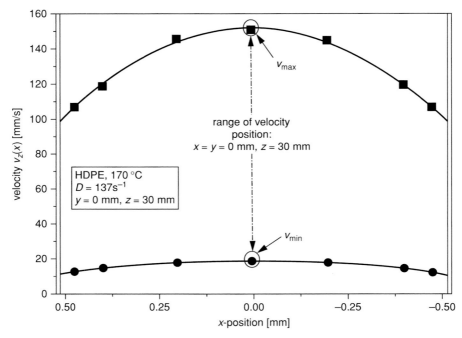

Figure 12.13. Laser-Doppler velocity profiles for a HDPE melt at the high and low parts of the cycle. Reprinted with permission from Munstedt et al., *J. Rheol.*, **44**, 413 (2000).

upper branch. Figure 12.13 shows laser-Doppler velocimetry data by Munstedt and co-workers on an HDPE during the two parts of the cycle at a nominal shear rate of $137 \, \text{s}^{-1}$; the slip velocity on the upper branch is ten times that on the lower branch.

The fact that a nominally incompressible melt can flow at two very different rates when driven by a piston moving at a constant rate is initially quite surprising, but the resolution, which was first pointed out by J. R. A. Pearson, is actually quite straightforward. There is a large reservoir of melt in the cylinder prior to the contraction to the capillary where the pressure drop is measured and the flow curve is determined. The melt, like all organic liquids, has a very small degree of compressibility. Because of the very large difference in volume between the reservoir and the capillary, it turns out that the pressure difference between operations on the two branches is sufficient to compress the melt in the cylinder by a volume that is on the order of the volume of the capillary. Thus, the system has a very large capacitance on the scale of the capillary, and this capacitance is what permits the oscillations.

The presence of the capacitance and our knowledge of the relative magnitudes of the slip velocities on the two branches of the flow curve lead to a straightforward conceptual model for the oscillatory flow. We assume for simplicity that there is a large discontinuity in the slip velocity at a critical wall stress τ_c, rather like the behavior seen in Figure 12.6. At stresses below τ_c there is little or no slip, the flow is slow, and the pressure in the reservoir may build up and compress a small amount of melt. When the pressure builds to a value where the wall stress must exceed τ_c to maintain the balance of forces, the melt will begin to slip, the flow rate will increase

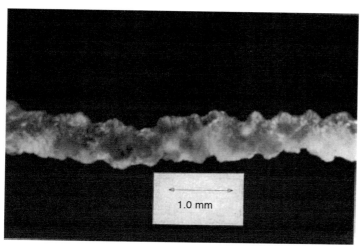

Figure 12.14. Extrudate of an LLDPE exhibiting melt fracture at an apparent shear rate of $6,655 \text{ s}^{-1}$. Reprinted from Kalika and Denn, *J. Rheol.*, **31**, 815 (1987).

greatly, the compressed melt will surge out of the reservoir, and the pressure will drop. When the pressure drops sufficiently the cycle starts over. The cycle time will depend linearly on the capacitance, hence, on the volume of melt in the reservoir, and this is in fact what is observed experimentally. This simple picture can of course be quantified, and a number of authors have done so. The specific response and dynamic flow rate and pressure profile cycle will depend on the details of the slip function and the stress-shear rate relation, but it is clear that cyclic behavior must result with nothing more than a capacitance and a "switch" at the wall.

The brass die data in Figure 12.2 do not show a discontinuity in the flow curve, but rather approach the high-rate slip-dominated behavior smoothly. These data illustrate the significant role that can be played by the interfacial chemistry, which determines the degree of polymer chain adsorption and hence stress transmission through the melt at the surface.

12.8 Melt Fracture

All entangled polymer melts, whether or not they exhibit sharkskin or slip-stick, are limited in throughput by the development of distorted extrudates like the LLDPE sample shown in Figure 12.14. Such high-rate extrudate distortions are gathered collectively under the name *melt fracture*; the term was coined by Tordella, who was one of the first to study the phenomenon, because he heard crackling sounds in the die that seemed to him to represent a real fracture of the polymer. Melt fracture for linear polymers that exhibit slip-stick typically occurs some distance out on the upper branch. Extrudates on the first part of the branch tend to be smooth, and there is some tradition of ultra-high speed processing by operating in the first portion of the upper branch; one-gallon high-density polyethylene milk bottles manufactured

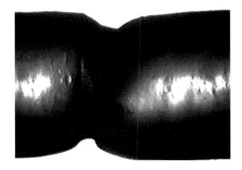

Figure 12.15. Apparent "weld" in extrusion of LLDPE. Reprinted with permission from Pérez-González and Denn, *Ind. Eng. Chem. Res.*, **40**, 4309 (2001). Copyright American Chemical Society.

with the Uniloy high-rate blow-molding process are processed in this regime, for example.

There have been attempts to explain the onset of melt fracture by using linear stability theory, along the lines discussed in the preceding chapter for the spinline. The procedure is the same in principle, but the computational problem to solve the relevant linear eigenvalue problem for channel flow with any viscoelastic constitutive equation is a very difficult one. Based on a number of successful solutions it appears that plane laminar flow of viscoelastic liquids at very low Reynolds numbers is stable to infinitesimal disturbances.

One possible mechanism for the onset of the extrudate instability is the propagation of a disturbance that originates in the die flow. There is a long folkloric tradition that melt fracture is an "entry" instability, and old flow visualization experiments showed the onset of a swirling and pulsating flow in the region upstream of a contraction that might correlate with a downstream instability. The problem here is one of determining cause and effect, since low Reynolds number flows can be strongly influenced by conditions far from the point of observation.

The question then is what might be the mechanism for an entry instability? One possibility is a tensile failure in the extensional flow in the die entry. Figure 12.15 shows a section of an extrudate of an LLDPE that appears to have been ruptured cohesively and then "welded" back together, which would be an indication of tensile failure, but such data are limited and the connection to a flow instability is tenuous. (The flow was stable when this sample was extruded.) Another possibility follows from the observation that noticeable slip occurs at stresses close to, and perhaps just below, the onset of extrusion instabilities. As noted in Section 12.5, slip in a converging region would lead to curved streamlines, and curved streamlines in highly elastic liquids appear to drive flow instabilities.

A related idea that would put the initiation of the instability in the die land rather than in the entry has recently been quantified. Linear stability theory does seem to predict in all cases that plane channel flow is stable to infinitesimal disturbances, but the time scale for perturbations to decay is a long one. That means that perturbed – that is, curved – streamlines will persist despite the linear stability of the base flow, and it is conceivable that this quasi-steady streamline curvature might be sufficient to drive an instability. The idea is still tenuous at the time of writing, however, and the mechanism for melt fracture is still unknown.

12.9 Concluding Remarks

There is a certain irony in the fact that the topics of wall slip and extrudate distortions, which have garnered so much attention in the periodical literature during the past two decades, receive so little attention in this text on modeling. The fact is, however, that our level of fundamental understanding of both of these related topics remains quite limited. Incorporation of wall slip in asymptotic analyses or simulations is usually straightforward, but a reliable and general slip law still does not exist, so any implementation will be ad hoc and without generality. Rules of thumb for the onset of extrusion instabilities have been essentially unchanged for four decades or more, and "processing aids" to avoid unstable regimes are readily available, so there is a certain complacency in applications based on the fact that one usually knows where the problems will occur and, if there is no processing aid that will fix the problem, the operating regime simply needs to be avoided.

This having been said, there are major unsettled theoretical issues regarding highly entangled polymer melts, and it is likely that the problems discussed in this chapter will be better understood when the fundamental issues in polymer physics have been resolved.

BIBLIOGRAPHICAL NOTES

There is an overview of wall slip and extrusion instabilities in polymer melts, with extensive references to earlier literature, in

Denn, M. M., "Extrusion Instabilities and Wall Slip," in J. L. Lumley et al., Eds. *Annual Review of Fluid Mechanics*, Vol. 33, Annual Reviews, Palo Alto, CA, 2001, p. 265.

Similar issues are addressed in an earlier chapter,

Denn, M. M., "Issues in Viscoelastic Fluid Mechanics," in J. L. Lumley et al., Eds. *Annual Review of Fluid Mechanics*, Vol. 22, Annual Reviews, Palo Alto, CA, 1990, p. 13.

The details of the tube theory for slip, with references to similar approaches using continuum formulations of tube theories, are in

Xu, F., M. M. Denn, and J. D. Schieber, *J. Rheol.*, **51**, 452 (2007).

The converging flow calculation with slip is in

Joshi, Y. M., and M. M. Denn, *J. Non-Newtonian Fluid Mech.*, **114**, 185 (2003).

There are extensive discussions of extrusion instabilities in the two *Annual Review of Fluid Mechanics* chapters cited at the beginning of this section. Comprehensive reviews that focus on the earlier literature are

Petrie, C. J. S., and M. M. Denn, *AIChE J.*, **22**, 209 (1976),
Larson, R. G., *Rheol. Acta*, **31**, 213 (1992),

as well as sections of

Pearson, J. R. A., *Mechanics of Polymer Processing*, Elsevier Applied Science Publishers, London, 1985.

A new book on the subject in which several chapters address the issues discussed here is

Hatzikiriakos, S. G., and K. Migler, Eds., *Polymer Processing Instabilities: Control and Understanding*, Marcel Dekker, New York, 2004.

For a description of melt fracture and the high-rate blow-molding process, see

Schaul, J. S., M. J. Hannon, and K. F. Wissbrun, *Trans. Soc. Rheol.*, **19**, 351 (1975).

The definitive experiments showing that sharkskin is a rupture phenomenon at the point of departure from the die are in

Shaw M. T., and L. Wang, *Proc. XIIIth Congress on Rheology*, Cambridge, UK, vol. 3, 2000, p. 170.
Migler, K. B., Y. Song, F. Qiao, and K. Flynn, *J. Rheol.*, **46**, 383 (2002).
Mizunuma, H., and H. Takagi, *J. Rheol.*, **47**, 737 (2003).

Calculations coupled with experiments intended to demonstrate that the tensile stresses at the onset of sharkskin are comparable to those required for rupture are in

Rutgers, R., and M. Mackley, *J. Rheol.*, **44**, 1319 (2000).
Pol, H. V., Y. M. Joshi, P. S. Tapadia, A. K. Lele, and R. A. Mashelkar, *Ind. Eng. Chem. Res.*, **46**, 3048 (2007).

The notion that melt fracture might be a subcritical instability caused by the presence of curved streamlines is explored in

Muelenbroek, B., C. Strom, A. N. Morozov, and W. van Saarloos, *J. Non-Newtonian Fluid Mech.*, **116**, 235 (2004).
Morozov, A. N., and W. van Saarloos, *Phys. Rev. Lett.*, **95**, 024501 (2005).

There are supporting numerical calculations in

Atalik, K., and R. Keunings, *J. Non-Newtonian Fluid Mech.*, **102**, 299 (2002).

13 Structured Fluids

13.1 Introduction

Many polymeric liquids have a microstructure even at rest. This might be a consequence of the presence of dispersed particulates or, in the case of liquid crystalline polymers, because of the rigidity of the polymer molecules. Continuum equations describing the stress and microstructure evolution are available for some limiting cases, permitting calculations of flow in complex geometries. The levels of description of the stress states are not comparable to that for entangled flexible polymer melts, so the resulting calculations are less likely to be in quantitative agreement, but they are still very useful for gaining insight into the development of morphology. We address three cases of structured fluids in this chapter: fiber suspensions, such as those that might be used for thermoplastic composites; liquid crystalline polymers; and fluids that exhibit a yield stress, which might include nanoparticle-filled melts.

13.2 Fiber Suspensions

The continuum approach to the rheology of fiber suspensions is based on a 1922 solution by Jeffery for the creeping-flow mechanics of a single ellipsoid in a shear flow. The ellipsoid rotates in a nonsinusoidal fashion, spending most of the period near a fixed angle to the flow direction. The ellipsoid aligns with the flow direction at all times in the limit of an infinite aspect ratio. The key assumptions in deriving a constitutive equation for a fiber suspension from Jeffery's result for the ellipsoid are that the suspending fluid is Newtonian and the suspension is dilute.

Let $\mathbf{n}$ be a unit vector that describes the axis of an ellipsoid of rotation. The total stress is then derived from the Jeffery stresses on the ellipsoid in terms of the microstructural orientation distribution tensor $\mathbf{A} = <\mathbf{nn}>$ as

$$\sigma = -p\mathbf{I} + \eta_m \mathbf{D} + \phi \left[\mu_0 \mathbf{D} + \tfrac{1}{2}\mu_2 \mathbf{D}\colon <\mathbf{nnnn}> + \mu_3 (\mathbf{D}\cdot\mathbf{A} + \mathbf{A}\cdot\mathbf{D}) \right]. \quad (13.1)$$

Here, η_m is the viscosity of the Newtonian matrix fluid and $<\ldots>$ denotes an ensemble average. $\mathbf{D} = \nabla\mathbf{v} + \nabla\mathbf{v}^T$ and $\mathbf{\Omega} = \nabla\mathbf{v} - \nabla\mathbf{v}^T$. The microstructure orientation distribution tensor **A** evolves according to the equation

$$2\frac{D\mathbf{A}}{Dt} = \mathbf{A}\cdot\mathbf{\Omega} - \mathbf{\Omega}\cdot\mathbf{A} + \frac{r^2-1}{r^2+1}\left[\mathbf{D}\cdot\mathbf{A} + \mathbf{A}\cdot\mathbf{D} - 2\mathbf{D}\text{: <nnnn>}\right]. \tag{13.2}$$

Here, r is the aspect ratio of the ellipsoid. Equation 13.2 follows from the requirement that the ellipsoid rotate with the local fluid but that it retain a unit length.

These equations require the fourth-order moment <**nnnn**>, which is unknown; an equation can be derived for the fourth-order moment, but it would involve the sixth-order moment, and so forth. We therefore need to express <**nnnn**> in terms of **A** = <**nn**>. This is the standard closure problem that always arises in statistical mechanics. There have been a number of studies to derive the best closure, but for our purposes it suffices to take the simplest form and to write

$$\mathbf{D}\text{: <nnnn>} \approx (\mathbf{D{:}A})\mathbf{A}. \tag{13.3}$$

Equation 13.3 preserves the period of rotation of the underlying particles in shear flow, and it gives excellent agreement with the exact values of <**nn**> and <**nnnn**> in shear and in biaxial extension, but better closures for general flows are known.

The coefficients are all proportional to the matrix viscosity and they are known exactly in terms of the aspect ratio. We record them here only in the limit $r \to \infty$, which is the case of usual interest:

$$r \to \infty: \quad \mu_0 = 2\eta_m, \quad \mu_3 = 0, \quad \mu_2 = \eta_m r^2/\ln r. \tag{13.4}$$

Equation 13.4 is a very good approximation to the exact result for all $r \geq 10$. In the limit $r \to 1$ (a sphere), Equation 13.1 simplifies to $\boldsymbol{\sigma} = -p\mathbf{I} + \eta_m(1 + 2.5\phi)\mathbf{D}$; $\eta_m(1 + 2.5\phi)$ is Einstein's equation for the viscosity of a dilute suspension of spheres.

It is relatively straightforward in principle to implement Equations 13.1 and 13.2 in a finite element code, but there is an interesting theoretical issue that has very practical implications for computation. The derivation of the constitutive equation assumes that the fiber has an aspect ratio but that it has no excluded volume; the only manifestation of the structure is the orientation tensor field **A**. Now consider fully developed laminar flow in a channel. The velocity **v** must vanish at the wall, and there will be a fixed value of the velocity gradient at the wall. It then follows from Equation 13.2 that **A** takes on a fixed, constant value at the wall. But **n** follows a rotational Jeffery orbit for any finite aspect ratio, so **A** = <**nn**> will be periodic for a finite aspect ratio at all positions arbitrarily close to the wall, and the transition from particles freely rotating near the wall to a fixed orientation at the wall is singular. Numerical resolution of the arbitrarily large gradients arising from this singular behavior is not possible on any mesh. There is a similar singularity along the centerline. This problem would not exist, of course, if particles at a finite distance from the wall could "feel" the presence of the wall because of their excluded volume.

There are ad hoc ways to deal with this singularity, which exists in all constitutive theories of this type. One approach, which complicates the programming considerably, is to use finite elements to solve for the velocity field but to integrate

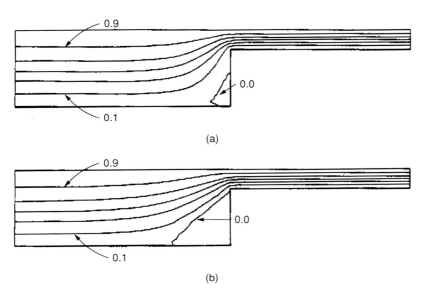

Figure 13.1. Computed streamlines for flow through a 4:1 contraction for (a) a Newtonian fluid ($\phi\mu_2/\eta_m = 0$) and (b) $\phi\mu_2/\eta_m = 5$. Reprinted from Lipscomb et al., *J. Non-Newtonian Fluid Mech.*, **26**, 297 (1988).

Equation 13.2 directly along individual streamlines, thus avoiding the wall problem completely but necessitating considerable interpolation in moving between the two methodologies for different parts of the coupled equations. Another is to employ the *aligned fiber approximation*, which makes use of the fact that $\mathbf{A} = \mathbf{vv}/|\mathbf{v}\cdot\mathbf{v}|$ is a stable solution to Equation 13.2 for infinite fibers under certain well-defined conditions regarding the velocity field. The procedure is then to use this approximation for $\mathbf{A}$ in Equation 13.1 but to compute the coefficients for the stress using the actual aspect ratio.

Figure 13.1 shows a calculation of streamlines for flow through a 4:1 contraction using the aligned fiber approximation and the coefficients for an infinite fiber (Equation 13.4). Figure 13.1a is for a Newtonian fluid ($\phi\mu_2/\eta_m = 0$), and Figure 13.1b is for a suspension with $\phi\mu_2/\eta_m = 5$ ($\phi = 5\ln r/r^2$, or $\phi \sim 0.2$ vol% for $r = 100$). The most striking feature of the suspension flow is the large increase in the size of the recirculating corner vortex. This calculation illustrates an important feature of the behavior of structured liquids: A small change in the structure can have a significant effect on the flow field. It is often tempting to uncouple the structure from the flow and to assume that the kinematics can be computed for the matrix fluid alone, after which the stresses and orientation distribution can be computed using the computed flow field. Such an approach (which can often be found in the fiber suspension literature) would lead to a major error in this case.

Figure 13.2 shows experimental streamlines for the flow of a 0.045-vol% suspension of 3.2-mm (1/8″) chopped glass fibers ($r = 276$) suspended in a corn syrup with a viscosity of about 20 Pa s through a 4.5:1 contraction. The downstream tube had a diameter of 5.6 mm. The computation in the left side of the picture was carried

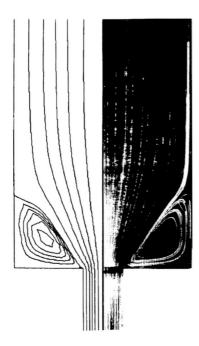

Figure 13.2. Comparison between computed and experimental streamlines in a 4.5:1 contraction, 0.045% (vol) 3-mm glass fibers in corn syrup. Reprinted from Lipscomb et al., *J. Non-Newtonian Fluid Mech.*, **26**, 297 (1988).

out using a finite element code with the aligned fiber approximation. The agreement between the computed and experimental flow patterns is quite good. The presence of the fibers clearly has a substantial effect on the flow. The size of the recirculating corner vortex is an important indicator: The reattachment point is at 0.18 upstream column diameters for the Newtonian suspending fluid and at 0.55 diameters for the suspension.

The dilute regime requires that fibers be noninteracting. Geometrical arguments show that this regime requires that ϕr^2 be small compared to unity, which is a criterion that is violated by both sets of calculations shown here despite the good agreement with experiment in Figure 13.2. It can be shown that the same form of the equations, but with a value for μ_2/η_m that depends on the actual dimensions of the particles and the mean spacing, is valid in the semidilute regime, where ϕr must be small compared to unity. This form of the equation has been used by a number of authors, usually assuming that the mean spacing is that of a random orientation of fibers.

Efforts to address highly concentrated systems have been of two types. One has utilized direct simulation of particle motion and interactions. Results from this approach are instructive, but it does not lead directly to continuum constitutive equations that can be used for simulation in complex flow geometries. The other approach has been to try to extend the dilute suspension theory into the concentrated regime by introducing irreversibility, which is an important characteristic of concentrated fiber suspensions. Folgar and Tucker pioneered this approach in 1984 by adding an ad hoc diffusion term to the equation for the orientation distribution, based on the notion that fiber interactions cause dispersion in a manner that is analogous to diffusion at the molecular level. These equations have been used with some success for mold filling simulations of fiber-filled polymer composites. The apparent diffusion coefficient is a phenomenological parameter that is generally taken to be proportional to the magnitude of the deformation rate. Phan-Thien and co-workers recently fit the results of particle simulations to obtain an empirical equation for the dependence of the diffusion coefficient on ϕr.

13.3 Liquid Crystalline Polmers

Liquid crystals are materials that flow like liquids but are ordered in the fluid state. They are typically made up of rigid molecules with a large aspect ratio and functional groups that are responsive to electromagnetic fields. *Nematic* liquid crystals have orientational order but no positional order (i.e., the centers of mass are randomly distributed, but the molecules retain a preferred orientation). Because the order is responsive to electromagnetic fields, these materials have found extensive use in display technology.

Some polymers exist in a nematic state. Nematic polymers have high moduli and high use temperatures, and they are resistant to chemical attack. Because of the molecular rigidity it is relatively easy in an extensional flow to cause these polymers to have an extended-chain morphology, so they can be processed as fibers with a very high tenacity. The first such commercial material was poly(p-phenylene terephthalamide), sold commercially as Kevlar® and Twaron®. Poly(p-phenylene terephthalamide) decomposes before it melts, and it must be spun in a gel-like state in a strong acid solvent. Several liquid crystalline polyesters that are marketed commercially are processed in the melt (*thermotropic* liquid crystalline polymers) and are used for specialized molding applications; they are not used extensively because they are expensive to produce and because they form micron-scale domain structures in molded parts that cause a loss in properties. The domain texture can clearly be seen in Figure 13.3, which is a polarized micrograph of Vectra B, a random copoly(ester amide) consisting of a 3:1:1 molar ratio of 6-hydroxy-2-naphthoic acid, terephthalic acid, and 4-aminophenol. This texture cannot be removed, even with the application of strong fields.

The first rheological theory for liquid crystals was developed by Leslie and Ericksen, building on Ericksen's earlier *transversely isotropic fluid*. The theory is formulated in terms of a director field **n**, and it is similar to the fiber theory in the preceding section, except that it includes a contribution to the free energy from an interactive potential that causes the molecules to align at rest. The usual form of the free energy F resulting from distortions of the director field is

$$2F = K_{11}\left(\nabla \cdot \mathbf{n}\right) + K_{22}\left(\mathbf{n} \cdot \nabla \times \mathbf{n}\right) + K_{33}|\mathbf{n} \times \nabla \times \mathbf{n}|^2, \qquad (13.5)$$

where K_{11}, K_{22}, and K_{33} are *Frank elastic constants* corresponding to splay, bend, and twist deformations of the director field, respectively. The stress

Figure 13.3. Vectra B observed under crossed polarizers after shearing at a stress of 500 Pa for 500 strain units. Reprinted from E. G. Kim, M.S. thesis, U. California, Berkeley, 1996.

equation is then written

$$\sigma = -p\mathbf{I} - \frac{\partial F}{\partial \nabla \mathbf{n}} \cdot \nabla \mathbf{n}^T + \tfrac{1}{2}\alpha_1 \mathbf{D}{:}\mathbf{nnnn} + \alpha_2 \mathbf{n}{\cdot}\mathbf{N} + \alpha_3 \mathbf{N}{\cdot}\mathbf{n} + \tfrac{1}{2}\alpha_4 \mathbf{D}$$

$$+ \tfrac{1}{2}\alpha_5 \mathbf{nn}{\cdot}\mathbf{D} + \tfrac{1}{2}\alpha_6 \mathbf{D}{\cdot}\mathbf{nn}, \tag{13.6a}$$

$$\mathbf{N} \equiv \frac{D\mathbf{n}}{Dt} + \tfrac{1}{2}\boldsymbol{\Omega}{\cdot}\mathbf{n}. \tag{13.6b}$$

Note that the contribution to the stress from the nematic potential is independent of the deformation rate and is therefore elastic. The coefficients $\{\alpha_i\}$ are known as the *Leslie viscosities*. (The factors of $\tfrac{1}{2}$ in the equations do not appear in the original literature because of different definitions of $\mathbf{D}$ and $\boldsymbol{\Omega}$.) The Onsager reciprocal relations from irreversible thermodynamics require that $\alpha_2 + \alpha_3 = \alpha_6 - \alpha_5$. Conservation of angular momentum must also be satisfied by the director, which takes the form

$$\mathbf{n} \times \left[\frac{\partial F}{\partial \mathbf{n}} - \nabla \cdot \left(\frac{\partial F}{\partial \nabla \mathbf{n}} \right) + (\alpha_3 - \alpha_2)\mathbf{N} + \tfrac{1}{2}(\alpha_3 + \alpha_2)\mathbf{D}{\cdot}\mathbf{n} \right] = 0. \tag{13.6c}$$

The director takes on a specified orientation at the boundary. The strength of the nematic effect in flow is determined by a dimensionless *Ericksen number* $E = (\alpha_3 - \alpha_2)L^2\dot{\gamma}/K$, where L is the characteristic length scale and K is a representative Frank elastic constant. Orientation boundary layers will develop because of competition between the alignment induced by the preferred boundary orientation and the nematic potential in the bulk.

Figure 13.4 shows a finite difference solution of the developing flow in a plane channel for a nematic liquid with the physical properties of 4'-octyl-4-biphenylcarbonitrile (8CB) at 34 °C, as given in Table 13.1. (K_{22} does not enter into this two-dimensional calculation.) The wall is at $y = 0$ and the midplane is at $y = 0.5$. The director surface orientation is taken to be parallel to the surface. E was varied from 10 to 70. A rather complex texture develops. This is a consequence of the tendency of the director to rotate, just like the ellipsoids in the Jeffery theory,

Table 13.1. *Nematic coefficients of 8CB at 34 °C*

α_2 (Pa s)	α_4	α_5	α_6	$K_{11}(N)$	K_{33}
−7.1	6.0	6.3	−3.2	1.41×10^{11}	2.09×10^{11}

but with an opposing force caused by the long-range effect of surface anchoring through the nematic potential. The sequence of textural transitions with increasing E is sometimes known as the *Ericksen number cascade*.

The Leslie-Ericksen theory appears to work well in describing the mechanics of low molar-mass liquid crystals, but it is too restrictive to describe the complex mechanics of liquid crystalline polymers. In particular, the theory cannot predict complex textures like the one in Figure 13.3, where the length scale of the texture is independent of the scale of the flow geometry. The free energy minimum should be a monodomain, but thermotropic liquid crystalline polymers appear to have a glasslike energy landscape with deep minima that retain texture. These submicron defect structures are not well understood, but they are believed to play a major role in the rheology at low and intermediate stresses.

Doi developed a constitutive theory for liquid crystalline polymers that takes into account the rotational diffusion of the large rodlike molecules and reduces to the Leslie-Ericksen theory for slow deformation rates, and this has been generalized by Marrucci and Greco and others to incorporate a nematic potential that accounts for gradients in the orientation tensor. The theory has a structure that is

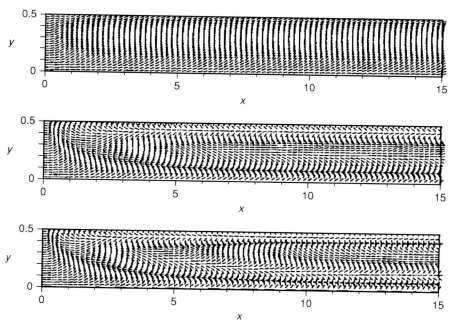

Figure 13.4. Developing flow between parallel planes with parameters of 8CB. $E = 10$ (top), 50 (middle), and 70 (bottom). Reprinted from Chono et al., *J. Non-Newtonian Fluid Mech.*, **79**, 515 (1998).

of the general form of Equations 13.1 and 13.2, but it is considerably more complex in detail because of the more complex physics that it is attempting to represent, and it includes terms involving $\nabla^2 A$. The only applications to date have been to simple shear flow, where complex textures with length scales that are independent of the macroscopic scale of the flow have been observed.

13.4 Yield-Stress Fluids

Many complex fluids exhibit a yield stress, in which the response is solidlike below a critical stress level and fluidlike above. Foods and consumer products often have a yield stress, as do some block and graft copolymer melts. Polymer melts with dispersed colloidal or nanoscale particles above a concentration sufficient to form a connected structure are likely to exhibit a yield stress, and such materials are of growing interest with the development of methods for the dispersal of nanoscale particles like exfoliated clays and carbon nanotubes.

The rheology of yield-stress (or *viscoplastic*) fluids is complex and often time dependent. Considerable insight can be gained, however, by considering the simplest example, the *Bingham material*. The classical Bingham material is defined for a shear flow with a positive shear rate as

$$\tau = \tau_y + \eta_p \dot{\gamma}, \quad \tau \geq \tau_y; \tag{13.7a}$$

$$\dot{\gamma} = 0, \quad \tau < \tau_y. \tag{13.7b}$$

There are two parameters, the yield stress τ_y and the "plastic viscosity" η_p. (Plastic viscosity is a terrible name for this parameter, but it is firmly embedded in the literature. While η_p has the dimensions of a viscosity, it is not the viscosity of the material while it is flowing. The true viscosity during flow is $\eta = \eta_p + \tau_y/\dot{\gamma}$, which is highly shear thinning.)

Oldroyd presented a properly invariant form of the extra-stress of a Bingham material in 1947 that also permits elastic deformation below the yield stress:

$$\tau = \left[\frac{\tau_y}{(\frac{1}{2}\mathbf{D}:\mathbf{D})^{1/2}} + \eta_p\right]\mathbf{D}, \quad \frac{1}{2}\tau:\tau \geq \tau_y^2; \tag{13.8a}$$

$$\tau = \mathbf{GE}, \quad \frac{1}{2}\tau:\tau < \tau_y^2. \tag{13.8b_1}$$

E is the strain tensor. It is usually assumed that the material is inelastic prior to yielding, in which case $G \to \infty$ and Equation 13.8b$_1$ is replaced by

$$\mathbf{D} = 0, \quad \frac{1}{2}\tau:\tau < \tau_y^2. \tag{13.8b_2}$$

These equations admit the possibility of *yield surfaces* when the von Mises yield criterion $\frac{1}{2}\tau:\tau = \tau_y^2$ is satisfied. An important consequence of Equations 13.8b1–2 in either form is that *the only admissible kinematics in regions where the yield criterion is not satisfied are solid-body motions* (rotation and translation).

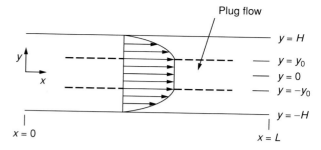

Figure 13.5. Pressure-driven flow of a Bingham material between parallel planes.

Consider fully developed pressure-driven laminar flow in a plane channel, with $v_x = v_x(y)$, $v_y = v_z = 0$. Following the development in Section 3.2.1, which is valid for any fluid, we find that the pressure gradient is a constant and the shear stress varies linearly across the channel, passing through zero at the center plane:

$$\tau_{xy} = -(\Delta p/L)y. \qquad (13.9a)$$

The stress then exceeds the yield stress only for

$$|y| \geq y_0 = \tau_y L/\Delta p. \qquad (13.9b)$$

There is no flow if y_0 is greater than the channel half-width H. We now integrate the x component of the momentum equation over $y_0 \leq |y| \leq H$ and apply the no-slip condition at the wall to obtain

$$v_x = -\frac{\Delta p}{2\eta_p L}(y^2 - H^2) + \frac{\tau_y}{\eta_p}(|y| - H), \quad y_0 \leq |y| \leq H. \qquad (13.10a)$$

Continuity of the velocity at $y = \pm y_0$ establishes the uniform velocity in the central undeformed core:

$$v_x = -\frac{\Delta p}{2\eta_p L}(y_0^2 - H^2) + \frac{\tau_y}{\eta_p}(y_0 - H), \quad |y| \leq y_0. \qquad (13.10b)$$

The velocity profile is shown in Figure 13.5. The logical sequence for obtaining this solution is important. The location of the region of solid-body motion is determined by the stress condition, which defines the equation for a surface. The magnitude of the solid-body motion cannot be specified, however; to do so would overdetermine the problem.

If we want to consider flows of yield-stress fluids in more complex geometries, we might expect that we could follow the development in Chapter 5 and apply the thin gap approximation. A bit of reflection quickly shows that this approach is not possible. Consider Figure 13.6, with a pressure-driven flow between nonparallel planes. According to the thin gap approximation, the velocity profile at each position in the flow direction would have the same shape, but the average velocity would decrease. The gap width H is now a function of position, in which case it follows from the preceding analysis that the velocity of the plug region must decrease as the channel width expands. But this is impossible if the plug is moving as a solid body. A similar conclusion follows by considering the radial flow in a center-gated disk mold, Section 6.2, shown in Figure 13.7. The unyielded region must expand, and

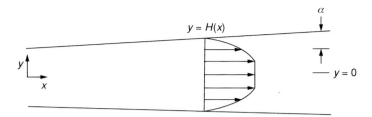

Figure 13.6. Pressure-driven flow between diverging planes. (Not a possible solution as shown.)

this is not possible if the region flows like an unyielded plug. Both of these arguments can readily be made quantitative, but this intuitive treatment should suffice.

There are two schools of thought about the consequences of the above reasoning. One holds that the concept of a yield stress is simply a convenient abstraction to represent a rapid change in properties as the shear rate goes to zero, and that the results should not be taken so literally. In support of this position we consider a fluid with a viscosity function like that shown in Figure 13.8; this fluid, sometimes known as the *biviscosity model*, appears to approach the Bingham fluid in the limit as $\eta_o \to \infty$, and it is unlikely that we could distinguish experimentally between a fluid with this rheology and a true yield-stress fluid. We *can* apply the thin gap approximation to such a purely viscous liquid, and the correct results for the two flows shown schematically in Figures 13.6 and 13.7 differ from the incorrect solution for a Bingham fluid only by terms of order η_p/η_o. The other school holds that the yield stress is a real material property, from which we would have to conclude that true unyielded regions are forbidden in most complex geometries. The essential difference in actual consequences between these two perspectives applies only to long-time static situations, in which the biviscosity liquid will deform continuously under any static stress, while the true yield-stress fluid will not.

The discontinuity in physical properties at a yield surface whose location is not known a priori makes numerical calculation in complex geometries extremely difficult with the exact Bingham material. The usual computational procedure is to use a regularization method, in which the Bingham material is approximated by a purely viscous fluid that goes to the true Bingham material in the limit as a small parameter goes to zero. The biviscosity fluid in Figure 13.8 is such a regularization, but

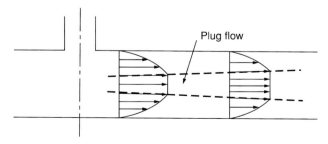

Figure 13.7. Radial flow in a center-gated disk mold. (Not a possible solution as shown.)

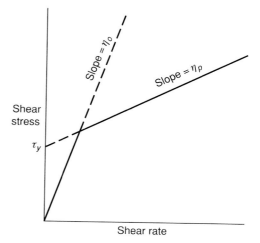

Figure 13.8. Shear stress-shear rate function of the biviscosity model to approximate a Bingham fluid.

it is more common to use smooth functions. This approach is intuitively appealing, but the limiting process is very delicate and true convergence to the discontinuous Bingham limit has been demonstrated only in restricted cases.

Two smooth regularizations that are frequently employed are

$$\boldsymbol{\tau} = \left[\frac{\tau_y}{\sqrt{\frac{1}{2}\mathbf{D}:\mathbf{D} + \varepsilon^2}} + \eta_p \right] \mathbf{D} \qquad (13.11a)$$

and

$$\boldsymbol{\tau} = \left[\frac{\tau_y \left[1 - \exp\left(-\varepsilon^{-1}\sqrt{\frac{1}{2}\mathbf{D}:\mathbf{D}} \right) \right]}{\sqrt{\frac{1}{2}\mathbf{D}:\mathbf{D}}} + \eta_p \right] \mathbf{D}. \qquad (13.11b)$$

It is impossible to go to the limit $\varepsilon \to 0$ with any finite grid, so there will always be a small amount of flow everywhere in the computational regime. Establishing convergence to a yield surface is therefore a difficult task (although some would argue that rigorous convergence does not matter, since the discontinuity in the constitutive equation itself is only an approximation).

Figure 13.9 shows the creeping flow of a Bingham fluid in a channel with an off-center disk using Equation 13.11a for the rheological model. This is the flow geometry that was employed in Figure 10.4 to represent the flow in a mold with a circular

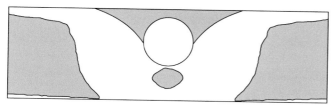

Figure 13.9. Flow of a Bingham material in a planar channel with an eccentric cylinder, $Bn = 125$. (Calculation by J. P. Singh.)

insert, and the geometric parameters are the same. Flows of Bingham materials are usually characterized in terms of the *Bingham number*, $Bn = \tau_y L / \eta_p V$. In this case the characteristic length is the radius of the cylinder, which is one-fourth the channel width, and the characteristic velocity is the mean upstream velocity. $Bn = 125$ for the example shown here. The shaded regions are unyielded, and the von Mises yield criterion, $\frac{1}{2} \boldsymbol{\tau} : \boldsymbol{\tau} = \tau_y^2$, is satisfied on the boundaries. Unlike the viscoelastic fluid shown in Figure 10.4, there is fore–aft symmetry for the Bingham fluid.

Away from the obstruction, the flow is like that for the fully developed channel flow in Figure 13.5, with a uniform plug that extends over most of the channel width. The flow has already started to adjust to the presence of the obstruction at the far left of the figure, causing the slight asymmetry. There is no flow in the smaller gap between the upper surface of the cylinder and the upper wall; for a plane channel this would correspond to a case in which y_0 in Equation 13.9b exceeds the channel half-width. The unyielded region in the larger space between the cylinder and the lower wall flows with a uniform velocity in the stream direction and has no transverse velocity; this is the region of maximum velocity.

Real yield-stress fluids typically have nonlinear stress-shear rate relations after yielding, and straightforward modifications to the procedures employed for the Bingham fluid are employed. The most common approach is to use an invariant form of the *Herschel-Bulkley* model, where the viscous term in the Bingham model is replaced by a power law. The presence of a yield stress (or even a close rheological approximation thereof) causes major changes in the deformation pattern of a fluid. One issue of practical relevance, for example, has to do with degassing; small bubbles may not have a sufficiently large buoyant force to overcome the integrated yield stress in a quiescent yield-stress fluid.

BIBLIOGRAPHICAL NOTES

The continuum formulation for fiber suspensions and the results shown here are from

Lipscomb, G. G., II, M. M. Denn, D. U. Hur, and D. V. Boger, *J. Non-Newtonian Fluid Mech.*, **26**, 297 (1988),

where references to earlier literature may be found. The definitions of **D** and $\boldsymbol{\Omega}$ in this article are different from those used here. Direct integration of the orientation distribution equations along streamlines is addressed in

Rosenberg, J., M. Denn, and R. Keunings, *J. Non-Newtonian Fluid Mech.*, **37**, 317 (1990).

The development of the diffusive term in the continuum equation for interacting fibers is in

Folgar, F. P., and C. L. Tucker, III, *J. Reinforc. Plast. Compos.*, **3**, 98 (1984).

A correlation for the coefficient of the diffusive term is in

Phan-Thien, N., X. J. Fan, R. I. Tanner, and R. Zheng, *J. Non-Newtonian Fluid Mech.*, **103**, 251 (2002).

An example of a simulation of fiber orientation development during mold filling, with references to earlier work, can be found in

Dou, H.-S., B. C. Cheong, N. Phan-Thien, K. S. Yeo., and R. Zheng, *Rheol. Acta*, **46**, 427 (2007).

The effect of various closure approximations on the prediction of mechanical properties of a fiber-filled composite is addressed in

Jack, D. A., and D. E. Smith, *Compos., Part A: Appl. Sci. Manuf.*, **38**, 975 (2007).

The literature on liquid crystalline polymers, where the same closure issues arise, has explored closure approximations beyond those studied in the fiber suspension literature; see

Chaubal, C. V., and L. G. Leal, *J. Rheol.*, **42**, 177 (1998).
Feng, J., C. V. Chaubal, and L. G. Leal, *J. Rheol.*, **42**, 1095 (1998).

The mechanics of liquid crystals are surveyed in

Rey, A. D., and M. M. Denn, *Ann. Rev. Fluid Mech.*, **34**, 233 (2002).

The classic text is

De Gennes, P. G., and J. Prost, *The Physics of Liquid Crystals*, 2nd ed., Oxford University Press, Oxford, 1993.

See also

Kleman, M., and O. D. Lavrentovich, *Soft Matter Physics*, Springer, New York, 2001.

The calculation in Figure 13.4 is from

Chono, S., T. Tsuji, and M. M. Denn, *J. Non-Newtonian Fluid Mech.*, **79**, 515 (1998).

The most detailed calculations with the Doi-Marrucci-Greco formulation for liquid crystalline polymers, which show the evolution of complex textures, are in

Klein, D. H., L. G. Leal, C. J. Garcia-Cervera, and H. D. Ceniceros, *Phys. Fluids*, **19**, 023101 (2007).

There is an extensive review of yield stress fluids in

Barnes, H. A., *J. Non-Newtonian Fluid Mech.*, **81**, 133 (1999).

Barnes is a leading proponent of the idea that the yield stress is not a material property. Yield stress measurement is addressed in

Nguyen, Q. D., and D. V. Boger, *Ann. Rev. Fluid Mech.*, **24**, 47 (1992).

There is a nice discussion of measurement issues in

Møller, P. C. F., J. Mewis, and D. Bonn, *Soft Matter*, **2**, 274 (2006).

A description of a triblock polymer melt that exhibits a yield stress is in

Hanson, P. J., and M. C. Williams, *Polym. Eng. Sci.*, **27**, 586 (1987).

For some examples of filled polymer melts that exhibit yield stresses with carbon nanotube, talc, and laponite clay fillers, respectively, see

Hobbie, E. K., and D. J. Fry, *J. Chem. Phys.*, **126**, 124907 (2007),
Kim, K. J., and J. L. White, *Polym. Eng. Sci.*, **39**, 2189 (1999),
Loiseau, A., and J.-F. Tassin, *Macromolecules*, **39**, 9185 (2006).

The development of plane channel flow and the thin gap approximation follows

Lipscomb, G. G., and M. M. Denn, *J. Non-Newtonian Fluid Mech.*, **14**, 337 (1984).

The regularizations in Equations 13.8a and b were introduced, respectively, by

Bercovier, M., and M. Engelman, *J. Comput. Phys.*, **36**, 313 (1980).
Papanastasiou, T. C., *J. Rheol.*, **31**, 385 (1987).

The biviscosity model seems to have been introduced simultaneously by Lipscomb and Denn, above, and

O'Donovan, E. J., and R. I. Tanner, *J. Non-Newtonian Fluid Mech.*, **14**, 75 (1984).

There is a detailed discussion of regularization methods, with references to the prior literature, in

Frigaard, I. A., and C. Nouar, *J. Non-Newtonian Fluid Mech.*, **127**, 1 (2005).

There is an interesting study of convergence of regularization in a complex flow for which the solution is known exactly in

Burgos, G. R., A. N. Alexandrov, and V. Entov, *J. Rheol.*, **43**, 463 (1999).

Much of the effort on complex flows of Bingham fluids has addressed unbounded flows past cylinders and spheres. The "gold standard" calculation of this type is

Beris, A. N., J. A. Tsamopoulos, R. C. Armstrong, and R. A. Brown, *J. Fluid Mech.*, **158**, 219 (1985).

There is a detailed look at convergence of the yield surface for flow past a sphere in

Liu, B. T., S. J. Muller, and M. M. Denn, *J. Non-Newtonian Fluid Mech.*, **102**, 179 (2002).

Squeeze flow of a Bingham fluid is studied in depth, with attention to convergence and the development of unyielded regions around stagnation points, in

Smyrnaios, D. N., and J. A. Tsamopoulos, *J. Non-Newtonian Fluid Mech.*, **100**, 165 (2001).

For subsequent work, see

Florides, G. C., A. A. Alexandrou, and G. C. Georgiou, *J. Non-Newtonian Fluid Mech.*, **143**, 38 (2007).

Flow of Bingham and Herschel-Bulkley fluids through a contraction, using the biviscosity model, is treated in

Jay, P., A. Magnin, and J. M. Piau, *J. Fluids Eng.*, **124**, 700 (2002).

For flow in an expansion, see

Mitsoulis, E., and R. R. Huilgol, *J. Non-Newtonian Fluid Mech.*, **122**, 45 (2004).

14 Mixing and Dispersion

14.1 Laminar Mixing

Mixing and blending in polymer processing applications almost always takes place in the laminar regime. The basic idea in laminar mixing is straightforward: Adjacent laminae of dissimilar materials are stretched – let us say doubled in length – so that the thicknesses of the laminae are reduced by a factor of two. The stretched sections are then folded back to create a block of the same thickness as the original, but it now contains four lamina instead of the original two. This process is repeated, and the number of lamina grows as 2^N, where N is the number of stretching/folding steps, while the thicknesses decrease as 2^{-N}. This process is known as the *baker's transformation*, for it is precisely the sequence of steps that is carried out in kneading a loaf of bread. (Push down to stretch, then fold back, turn 90°, and repeat.) Laminar mixing is related to the theory of chaotic dynamical systems, and it has been widely studied in the context of dynamical systems since the late 1980s.

The implementation of this methodology in polymer processing long predates the development of the theoretical tools currently in use for analysis. Static mixers are commonly employed to effect the baker's transformation. A cutaway view of a Kenics static mixer is shown in Figure 14.1. This device consists of a series of helically twisted blades that divide the circular channel into two twisted semicircular ducts. The leading edges of successive blades are placed at an angle, usually 90°, to the trailing edges of the preceding blades so that the new cut splits each of the previous segments. Thus, stretching takes place in the flow segments and folding at the transitions. Other commercial devices have different configurations but effect the same transformation.

Recent years have seen considerable effort to quantify this and other mixing geometries. The difficult problem is not so much the solution of the flow field in the mixing element, but rather the tracking of the fluid elements that make up the exponentially growing interface as the mixture progresses through the mixing channel (or as time progresses in a simulation of a batch mixer). We shall not address these issues here because they are highly technical and would be a major diversion from our objectives; it suffices to say that progress has been made but the tracking

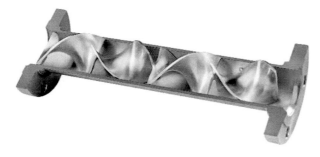

Figure 14.1. Cutaway view of a portion of a Kenics static mixer with two mixing segments.

problem is by no means solved for all cases of interest. Figure 14.2 shows a cross section from a three-dimensional finite element simulation of the flow of identical black and white, inelastic Carreau-Yasuda fluids with a power-law exponent of 0.1 through a static mixer with six blades with a twist angle of 180°. Calculations of this type can be used to optimize the design, which was one goal of this work, where the effect of the total twist angle of the blade was examined. The studies to date have been restricted to pairs of inelastic liquids with comparable viscosities and no interfacial tension. Neglect of viscoelasticity is probably not a serious shortcoming for this application, since most of the flow takes place through nearly parallel channels, but calculated pressure drops are likely to be affected by reorganization in the entry and exit regions of successive sections.

14.2 Droplet Breakup and Coalescense

Droplet breakup and coalescence are the primary physical processes in the mixing of liquids with very different viscosities. There is an extensive literature on the breakup of single droplets of Newtonian fluids in a Newtonian matrix, mostly building on a classic study by Grace that was first published in 1982 but was based on older work. Grace created a map of the critical capillary number $Ca = \eta_m \dot{\gamma} R / \sigma$

Figure 14.2. Simulation of the mixing pattern for two identical Carreau-Yasuda fluids with $n = 0.1$ after passing through a static mixer with six blades with a twist angle of 180°. Reprinted with permission from Galaktionov et al., *Int. Polym. Proc.*, **XVIII**, 138 (2003).

at breakup in a shear flow versus the viscosity ratio $p = \eta_d/\eta_m$, where subscripts d and m refer to dispersed and matrix phases, respectively. There is never breakup in shear if $p = \eta_d/\eta_m > 4$ or if Ca is less than 0.4. Breakup occurs by "tip streaming," where the droplet elongates and sheds small drops from the pointed ends, for $\eta_d/\eta_m \ll 1$, whereas breakup occurs by necking in the central portion of the droplet for $\eta_d/\eta_m = O(1)$. For Ca equal to about twice the critical value in shear, or five times the critical value in extension, the droplet elongates into a filament, and breakup occurs because of an interfacial tension-driven (*Rayleigh*) mechanism.

There have been a number of studies of the breakup of viscoelastic filaments because of interfacial tension, all of which show that the viscoelastic response results in dynamics that differ markedly from the Rayleigh breakup of inelastic filaments, but the focus for interfacial tension-driven breakup has been on dilute polymer solutions of interest in aerial spraying and inkjet technology applications. The literature is less extensive for breakup of polymer melts, and some of the issues relevant to dilute solutions are probably not of concern for melts because of the orders-of-magnitude difference in the ratio of interfacial to viscous (or elastic) stresses.

The primary computational issue for studying breakup or coalescence is tracking the interface. This has been done in a number of ways, none of which is clearly better than others. One approach is to track the interface explicitly and to evolve the mesh with time, remeshing if the elements become too distorted. Another approach is to employ the method of "level sets," where a scalar function $\varphi(\mathbf{x}, t)$ that is positive in one phase and negative in the other, with a value of zero on the interface, convects with the motion. The mass, momentum, and energy equations can then be formulated for the entire two-phase regime, with the parameters dependent on the algebraic sign of φ. (There are of course very important details regarding handling interfacial tension and ensuring that φ evolves in a way the conserves mass. These are addressed in specialized discussions.) The challenge in all methods is to control surface curvature continuity to ensure that spurious secondary flows driven by apparent interfacial forces caused by artificially large local curvature cannot arise. Finite element, volume of fluid, and boundary element methods have been used for solving the continuum equations. Most calculations have been done for individual Newtonian droplets.

Figure 14.3 shows a boundary element calculation of a sheared drop with adaptive discretization of the drop interface. The sequence of shapes is for a 50-μm, 100-Pa s polydimethylsiloxane droplet sheared in a 1-mm gap in a matrix of a polyisobutylene of nearly the same viscosity. The interfacial tension is 2.4 mN/m. The images are experimental data and the profiles are simulations using slightly adjusted parameters to ensure a good fit at pinch-off. The numbers are the experimental total strains (time multiplied by shear rate), while the numbers in parentheses are the computed total strains for the same droplet shapes. The agreement is quite good, although viscoelasticity would be expected to play a role in the dynamics of the pinch-off region. These calculations are typical of the state of the art at the time of writing.

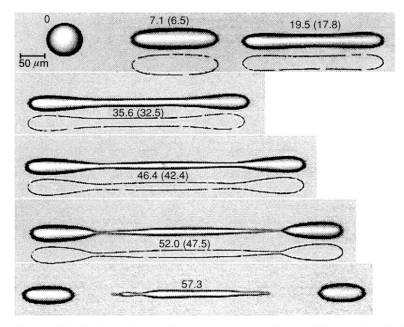

Figure 14.3. Deformation and breakup of a polydimethylsiloxane droplet in a polyisobuty-lene of nearly the same viscosity; viscosity $\approx$ 100 Pa s, interfacial tension $\approx$ 2.4 mN/m. Reprinted with permission from Cristini et al., *J. Rheol.*, **47**, 1283 (2003).

14.3 Continuum Stress Equations

Very good emulsion (*effective medium*) models that average over the continuous and dispersed phases to describe the continuum rheology have been developed to describe the linear viscoelasticity of polymer blends. The *Palierne* model, for example, provides a mixing rule for the complex modulus $G^* = G' + iG''$ in terms of the moduli for the dispersed and matrix phases as

$$G^* = G_m^* \frac{1 + 3 \sum \phi_i H_i^*}{1 - 2 \sum \phi_i H_i^*}, \qquad (14.1a)$$

$$H_i^* = \frac{4(\sigma/R_i)[2G_m^* + 5G_d^*] + [G_m^* - G_d^*][16G_m^* + 19G_d^*]}{40(\sigma/R_i)[G_m^* + G_d^*] + [2G_m^* + 3G_d^*][16G_m^* + 19G_d^*]}. \qquad (14.1b)$$

Here, ϕ_i is the volume fraction of droplets with radius R_i. The volume-averaged radius can be used instead of the distribution if the polydispersity is less than 2.3. The fit to G' data for a 6.5% dispersion of poly(ethylene terephthalate) in a linear fluoropolymer at 300 °C, with an average droplet radius of 5.4 μm, is shown in Figure 14.4, where the data are plotted as the ratio of the storage modulus of the blend to that of the matrix polymer. The peak reflects the frequency regime where the interfacial dynamical response is sensitive to interfacial tension. Interfacial tension acts like an elastic stress, so the major contribution of the interface is in G'. The interfacial tension of the blend estimated from standard equations is about

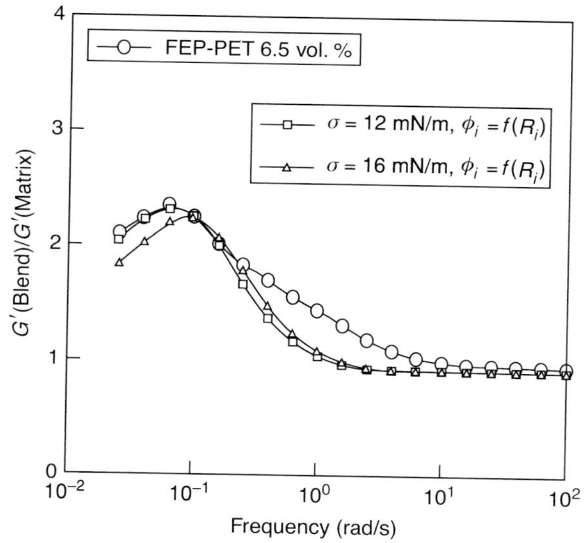

Figure 14.4. Scaled storage modulus of 6.5 vol% PET-FEP blend at 300 °C with theoretical curve using Equation 14.1. Reprinted from Lee and Denn, *J. Rheol.*, **43**, 1583 (1999).

16 mN/m, although the fit to the data is a bit better with $\sigma = 12$ mN/m. The theory is generally considered to be reliable enough to use as a means of measuring interfacial tension. It does not give good results when the dispersed phase is a liquid crystalline polymer.

The zero-shear viscosity of a dispersion can be extracted from Equation 14.1 by allowing the frequency to go to zero, giving

$$\eta_0 = \eta_{m0} \frac{100(1+p) + 3\phi(2+5p)}{10(1+p) - 2\phi(2+5p)} \approx \eta_{m0} \left[1 + \phi \frac{2+5p}{2(1+p)} + O(\phi^2) \right]. \qquad (14.2)$$

Blend theories for finite deformations have been developed from first principles only for two Newtonian phases. A theory of Choi and Schowalter, which is exact through first order in the deviation from a Newtonian droplet, gives

$$\eta = \eta_m \left[1 + \phi \frac{2(2+5p) - 10(p-1)\phi^{7/3}}{4(1+p) - 5(2+5p)\phi + 42\phi^{5/3} + 5(2-5p)\phi^{7/3} + 4(p-1)\phi^{10/3}} \right]$$
$$\approx \eta_m \left[1 + \phi \frac{2+5p}{2(1+p)} + O(\phi^2) \right]. \qquad (14.3)$$

The Palierne and Choi-Schowalter theories agree through first order in the volume fraction of the dispersed phase, but the quadratic term is considerably larger in the latter theory. Other theories and heuristics, including extensions to viscoelastic liquids, are discussed in a 2002 review by Tucker and Moldenaers. The rheology depends on the morphology of the dispersed phase, and the emphasis remains on understanding morphology and the mechanics of single droplets in viscoelastic systems. The continuum rheology of dispersed systems is therefore not well developed and remains a subject of current research.

14.4 Concluding Remarks

The intention of this brief chapter is to note the state of the art in two areas critical to many aspects of polymer processing, namely, laminar mixing and droplet dynamics. Except for the very important details of issues like interface tracking and interface smoothing, the computational methods used to attack these classes of problems are the same as those employed for the single-phase continuum calculations in the preceding chapters. There are some excellent analytical results for droplet breakup in the classical fluid mechanics literature, and they have provided a great deal of guidance for those interested in polymer processing. The development of useful continuum constitutive relations for the simulation of processing flows of dispersed systems is still a subject of research.

BIBLIOGRAPHICAL NOTES

The basic text on laminar mixing and the relation to chaotic dynamical systems is

J. M. Ottino, *The Kinematics of Mixing: Stretching, Chaos, and Transport*, Cambridge University Press, Cambridge, 1989.

The topics in this chapter are addressed broadly in two reviews,

Ottino, J. M., P. DeRoussel, S. Hansen, and D. V. Khakhar, *Adv. Chem. Eng.*, **25**, 105 (1999).
Tucker, C. L., III, and P. Moldenaers, *Ann. Rev. Fluid Mech.*, **34**, 177 (2002).

There is a nice introduction to the mechanics of static mixers, with some analytical results for pressure drops and other flow characteristics, in

Middleman, S., *Fundamentals of Polymer Processing*, McGraw-Hill, New York, 1977, pp. 327ff.

The mapping procedure for interface tracking in mixing is described in

Kruijt, P. G. M., O. S. Galaktionov, P. D. Anderson, G. W. M. Peters, and H. E. H. Meijer, *AIChE J.*, **47**, 1005 (2001).

There is a discussion of optimizing the static mixer in

Galaktionov, O. S., P. D. Anderson, G. W. M. Peters, and H. E. H. Meijer, *Int. Polym. Proc.*, **XVIII**, 138 (2003).

Recent work on quantifying mixing, which is a subject with a long history, can be found in

Camesasca, M., M. Kaufman, and I. Manas-Zloczower, *Macromol. Theory Simul.*, **15**, 595 (2006).

Interface tracking schemes for free-surface problems are reviewed in

Scardovelli, R., and S. Zaleski, *Ann. Rev. Fluid Mech.*, **31**, 567 (1999).

A finite element formalism is described in

Kistler, S. F., and L. E. Scriven, "Coating Flows," in J. R. A. Pearson and S. F. Richardson, Eds., *Computational Analysis of Polymer Processing*, Applied Science Publ., Barking, U. K., 1983, p. 243.

The level set method is described in

Sethian, J. A., and P. Smereka, *Ann. Rev. Fluid Mech.*, **35**, 347 (2003).

There are applications of level sets to bubble dynamics in Maxwell and Bingham fluids, respectively, in

Pillaipakkam, S. B., and P. Singh, *J. Comput. Phys.*, **174**, 552 (2001).

Singh, J. P., and M. M. Denn, *Phys. Fluids*, **20**, 040901 (2008).

The method is used to follow the front in mold filling of a fiber suspension in

Dou, H.-S., B. C. Cheong, N. Phan-Thien, K. S. Yeo, and R. Zheng, *Rheol. Acta*, **46**, 427 (2007).

The breakup calculations in Figure 14.3 are from

Cristini, V., S. Guido, A. Alfani, J. Blawzdziewicz, and M. Loewenberg, *J. Rheol.*, **47**, 1283 (2003).

The Palierne theory is developed in

Palierne, J. F., *Rheol. Acta*, **29**, 204 (1990); Erratum, **30**, 497 (1991).

The data in Figure 14.4 are from

Lee, H. S., and M. M. Denn, *J. Rheol.*, **43**, 1583 (1999).

Postface

This is the conclusion of the book, and an appropriate point to look back and reflect. Our goal throughout has been to establish the foundations of polymer melt processing in fluid mechanics and heat transfer without introducing unnecessary complexity. In doing so we have avoided geometrical detail of the equipment; such detail is important for specific applications, but its inclusion adds little to our overall understanding of the essential interplay between fluid mechanics and heat transfer in basic process performance, which was our primary objective. Similarly, we initially developed the subject in terms of the flow of inelastic liquids; many polymer processes are characterized by a low Deborah number, either as a consequence of the nature of the flow or the properties of the polymer being processed, and the essential behavior in this case does not depend on the fact that the melt is viscoelastic. We subsequently introduced viscoelasticity for those applications where it is needed, using viscoelastic constitutive equations that have been found to be effective in describing melt flow in complex geometries but fall short of the state of the art in polymer rheology. Viscoelasticity can be quite significant in some processing situations, notably in steady flows with substantial elongation and in all flows when dynamical response is of interest, and the rôle of viscoelasticity – when it is important and when it is not – must be understood for a complete and accurate picture.

At this point the reader should have the preparation necessary to continue with more focused texts and the periodical literature to address the important details about geometry and materials that have been bypassed here, and to go on to study other polymer processes and processes involving other complex liquids. The important periodical literature includes fundamental journals such as the *Journal of Rheology*, *Rheologica Acta*, the *Journal of Non-Newtonian Fluid Mechanics*, *Physics of Fluids*, *Macromolecules*, and the *Journal of Applied Polymer Science*, as well as journals that focus more on applications, such as *International Polymer Processing*, *Polymer Engineering and Science*, the *ANTEC Proceedings* of the Society of Plastics Engineers, and *Advances in Polymer Technology*. More general engineering journals, especially those directed to chemical and mechanical engineers, often have relevant articles.

We hope that this has been a fruitful endeavor that has motivated further study.

Author Index

Subject Index

aligned fiber approximation, 219
amplitude ratio, 128
    melt spinning and, for PET pilot data, 186, 188
    spinline sensitivity and, 188, 189
    in transfer function analysis, 184

Bagley plot, 149, 150
baker's transformation, 231
Bingham material, 224
    flow of, 227
    stress-shear rate functions for, 227
Biot number, 48, 100
blow molding, 10, 11, 214
blown film. *See* film blowing
blowup ratio, 163, 193
Bode diagram, 128
Boltzmann superposition integral, 134
boundary conditions, 30, 200
    for melt spinning, 92
    no-slip, 31, 199
branching, in polymer structures, 14

Carreau-Yasuda fluid, 35, 208
    laminar mixing for, 232
Cauchy momentum equation, 21, 23, 24
center-gated disk mold, 72
    coupled flow in, 76
    with isothermal Newtonian liquid, 72
    with isothermal power-law fluid, 75
    moving contact line in, 74
"choking," 56
    isothermal flow and, 56
    viscous dissipation and, 58
coating operations, 7
    sheet, 67, 114
Cole-Cole plot, 128
compression molding, 6, 77
    fiber orientation distribution during, 6
    "fountain flow" with, 81
    with isothermal Newtonian fluid, 77, 80

    with isothermal power-law fluid, 81
    liquid-air interface in, 79
conservation of energy, 29
    in spinline simulations, 107
conservation of linear momentum, 21
    Cauchy momentum equation in, 21
    in spinline simulations, 106
    stress and, 21
conservation of mass, 18, 21
    continuity equation in, 19
conservation principles, 18, 32
    boundary conditions and, 30, 32
    constitutive equations and, 24, 26
    for energy, 29
    for linear momentum, 21
    for mass, 18, 21
constitutive equations, in polymer flow, 24, 26, 141, 151
    Giesekus model, 148
    K-BKZ model, 149
    Maxwell model, 142, 144
    Phan-Thien/Tanner model, 145, 148
    reptation and, 148
    rheology and, 126
continuity equation, 19, 20
continuum stress equations, 234, 235
control volumes, 18
converging flow, 19, 21, 28, 165, 206
    fringe patterns for, 167
    stresses for, 167, 168
Cox-Merz rule, 139
creeping flow, for polymers, 26, 29
cross-head wire die, 7

Deborah number, 13, 153
differential scanning calorimetry (DSC), 16
Dissipation function, 30
draw ratios
    for Maxwell model, 158
    for Phan-Thien/Tanner model, 158, 182
draw resonance, 175, 182

Printed in the United States
By Bookmasters